39.35

In Eves' Circles

DATE DUE

In Eves' Circles

Joby Milo Anthony, Editor

MAA Notes Number 34

Published and Distributed by
The Mathematical Association of America

MAA Notes and Reports Series

The MAA Notes and Reports Series, started in 1982, addresses a broad range of topics and themes of interest to all who are involved with undergraduate mathematics. The volumes in this series are readable, informative, and useful, and help the mathematical community keep up with developments of importance to mathematics.

MAA Notes

1. Problem Solving in the Mathematics Curriculum, *Committee on the Teaching of Undergraduate Mathematics,* a subcommittee of the Committee on the Undergraduate Program in Mathematics, *Alan H. Schoenfeld,* Editor
2. Recommendations on the Mathematical Preparation of Teachers, *Committee on the Undergraduate Program in Mathematics, Panel on Teacher Training.*
3. Undergraduate Mathematics Education in the People's Republic of China, *Lynn A. Steen,* Editor.
5. American Perspectives on the Fifth International Congress on Mathematical Education, *Warren Page,* Editor.
6. Toward a Lean and Lively Calculus, *Ronald G. Douglas,* Editor.
8. Calculus for a New Century, *Lynn A. Steen,* Editor.
9. Computers and Mathematics: The Use of Computers in Undergraduate Instruction, *Committee on Computers in Mathematics Education, D. A. Smith, G. J. Porter, L. C. Leinbach, and R. H. Wenger,* Editors.
10. Guidelines for the Continuing Mathematical Education of Teachers, *Committee on the Mathematical Education of Teachers.*
11. Keys to Improved Instruction by Teaching Assistants and Part-Time Instructors, *Committee on Teaching Assistants and Part-Time Instructors, Bettye Anne Case,* Editor.
13. Reshaping College Mathematics, *Committee on the Undergraduate Program in Mathematics, Lynn A. Steen,* Editor.
14. Mathematical Writing, by *Donald E. Knuth, Tracy Larrabee, and Paul M. Roberts.*
15. Discrete Mathematics in the First Two Years, *Anthony Ralston,* Editor.
16. Using Writing to Teach Mathematics, *Andrew Sterrett,* Editor.
17. Priming the Calculus Pump: Innovations and Resources, *Committee on Calculus Reform and the First Two Years,* a subcomittee of the Committee on the Undergraduate Program in Mathematics, *Thomas W. Tucker,* Editor.
18. Models for Undergraduate Research in Mathematics, *Lester Senechal,* Editor.
19. Visualization in Teaching and Learning Mathematics, *Committee on Computers in Mathematics Education, Steve Cunningham and Walter S. Zimmermann,* Editors.
20. The Laboratory Approach to Teaching Calculus, *L. Carl Leinbach et al.,* Editors.
21. Perspectives on Contemporary Statistics, *David C. Hoaglin and David S. Moore,* Editors.
22. Heeding the Call for Change: Suggestions for Curricular Action, *Lynn A. Steen,* Editor.
23. Statistical Abstract of Undergraduate Programs in the Mathematical Sciences and Computer Science in the United States: 1990–91 CBMS Survey, *Donald J. Albers, Don O. Loftsgaarden, Donald C. Rung, and Ann E. Watkins.*
24. Symbolic Computation in Undergraduate Mathematics Education, *Zaven A. Karian,* Editor.
25. The Concept of Function: Aspects of Epistemology and Pedagogy, *Guershon Harel and Ed Dubinsky,* Editors.
26. Statistics for the Twenty-First Century, *Florence and Sheldon Gordon,* Editors.
27. Resources for Calculus Collection, Volume 1: Learning by Discovery: A Lab Manual for Calculus, *Anita E. Solow,* Editor.
28. Resources for Calculus Collection, Volume 2: Calculus Problems for a New Century, *Robert Fraga,* Editor.

MAA Reports

These volumes may be ordered from:
The Mathematical Association of America
1529 Eighteenth Street, NW
Washington, DC 20036
800-331-1MAA   FAX 202-265-2384

Howard Eves

by Joby Anthony

In 1979 Howard Eves retired from the University of Maine at Orono after a long and distinguished career. Howard grew up in New Jersey, demonstrating an aptitude for mathematics and a love of learning very early. In the eighth grade he read Cajori's *History of Mathematics* and Euclid's *Elements*. As a junior in high school he and a few friends studied Granville's *Calculus* in the evenings because they wanted to know what that mysterious subject was all about. After high school he entered the University of Virginia where he received a B.S. in mathematics He went on to Harvard for an M.A. in mathematics and then began work on a doctoral program with the well known geometer Julian Lowell Coolidge.

It was decided that a geometer should know tensor analysis, so Howard was sent to the Institute for Advanced Studies in Princeton to learn that subject. At the Institute he was assigned the responsibility for logistical details of the colloquium program. As part of this assignment he made the arrangements for Einstein's first public lecture in the United States. He formed a friendship with the older man, who on one occasion pointed out a small flower growing in a crack of the sidewalk, and told Howard to "...bloom where you are planted."

Howard's education was interrupted by the death of his father, and by the Great Depression. He left school to work as a surveyor of the Tennessee Valley Authority. But he really wanted to teach. He finally found a temporary job at Bethany College, and never again left academic life. He finished a Ph.D. at Oregon State University, and has subsequently taught and lectured at colleges and universities around the world.

Several generation of mathematicians and mathematics teachers have been enriched by Howard Eves. His "great moments in mathematics" books are among the most popular ever printed by the Mathematical Association of America and his mathematical circles books have delighted students and teachers for many years. Probably his most famous book is *An Introduction to the History of Mathematics* which has been a standard text for a history of mathematics course for over 40 years. Those of us who have known Professor Eves have called him "elvish," and indeed he is. At the age of 80 he was still teaching, lecturing, and producing mathematics with an almost magical (certainly elvish) quality. Howard has been a mathematician, writer, editor, and teacher, but perhaps the best description is found in a noun that is overworked and often misunderstood. He is, in the classical and very best sense of the word, a "scholar."

On May 9-11, 1991, the Department of Mathematics at the University of Central Florida honored Dr. Eves with a Conference on Geometry, Mathematics History, and Pedagogy. This conference celebrated his eightieth birthday. About 150 mathematicians and mathematics teachers attended the conference. This volume contains eighteen selected papers that were presented at the conference. There is something in these papers for almost everyone who is interested in mathematics. There are two papers covering "geometry, history, and pedagogy," two in "history and pedagogy," two in "problem solving," four in "geometry," four in "history," and three in "pedagogy." Howard has made significant contributions to all of these topics. It has been both a pleasure and a privilege for me to organize this conference in his honor, and to edit these proceedings.

Contents

Geometry, History, and Pedagogy

Professor Meserve begins this article with a short survey of the thirteen books of Euclid's *Elements*. He then uses Euclid to discuss the geometric representation of numbers and of algebraic concepts leading to the use of geometry to model the physical universe. He concludes with the roles of geometries in mathematics and science, and some objectives for the inclusion of geometry in the undergraduate mathematics curriculum.

In this paper Professors Hilton and Pedersen show how an approximation with any desired degree of accuracy to any regular convex N-gon (for example a regular pentagon) may be constructed from folding a straight strip of paper in a specific, repetitive pattern. In addition they show how the folding process for the regular convex N-gons may be modified to construct approximations to all of the regular star polygons. The paper-folding process leads to some surprising results in number theory.

History and Pedagogy

Professor Barbin has been deeply involved with the development of mathematics education programs in France for some time. In this paper she examines carefully the historical notion of "proof" in mathematics, and draws some conclusions about the role of proof for mathematics teachers. She addresses the question "Is proving a means to convince or to enlighten?" She concludes that when teaching "proof" the question "How do you prove this result?" should be replaced with "How have you obtained this result?"

This paper contains a brief history of some simple enciphering and deciphering techniques along with a collection of intriguing little "coding" problems. These problems could be used in a typical high school classroom as "enrichment" exercises that will absorb the attention of bright and inquisitive students.

Problem Solving

For many years Professor Dodge was a colleague of Howard Eves at the University of Maine where he assisted Howard with the Elementary Problem Department of the *American Mathematical Monthly*. Presently he is the editor of the problem section of the Pi Mu Epsilon Journal. This paper contains a collection of problems, mostly in geometry, that have challenged Professor Dodge over the years.

Professor Klamkin is the former coach of the American Mathematical Olympiad team. He presents a very well organized list of problem-solving strategies, and then applies them to a collection of fascinating problems that have appeared in various contests, books, and papers.

Geometry

A "pedal triangle" is formed in the following way. From a point P in the interior of a triangle, drop perpendicular line segments to the three sides of the triangle. The triangle whose vertices are the points of intersection of these line segments with the sides of the original triangle is called a "pedal" triangle. (A generalization of this concept produces the definition of a "pedal" n-gon.) A "bi-centric quadrilateral" is a quadrilateral with both a circumscribed and inscribed circle. Both bi-centric quadrilaterals and pedal n-gons have an extensive history. This paper uses advanced euclidean geometry to give some surprising results in both subject areas along with an interesting pedal n-gon conjecture and a new euclidean construction for the bi-centric quadrilateral.

Professor Peter Hilton is one of the best known of contemporary American mathematicians. This paper shows how techniques in algebraic topology can establish deep and beautiful links between geometry and algebra. Here are some comments by one of the referees. "It is not very difficult to learn a certain theory so that we are familiar with basic definitions and results and possibly aware of some applications. It is much more difficult to acquire a sufficient understanding of a theory so that we can effectively use its tools in new situations. The third stage is understanding how the studied theory fits within mathematics as a whole. This usually takes years of serious and devoted research and possibly some special qualities of mind. Most of us never really get there. Fortunately, it is possible to get glimpses of that world from mathematicians who have achieved that stage. This paper is in that category."

What is a "blue moon?" The answer may surprise you, and the mathematical connections between "blue" moons, pacemakers, and the study of crystals that don't quite have the symmetry that crystals are supposed to have (quasicrystals) will certainly surprise you. Professor Senechal shows how "circle" maps can be used to unify some aspects of the study of these apparently disparate phenomena.

A cubic curve is the locus of points in the plane that satisfy a third-degree polynomial equation. This paper in advanced euclidean geometry establishes some remarkable connections between ordinary triangles and families of cubic curves that are associated with that triangle.

History

There are only five "regular" solids; the tetrahedron, the cube, the octahedron, the dodecahedron, and the icosahedron. Three of these, the tetrahedron, the octahedron, and the icosahedron, have faces that are equilateral triangles. One, the cube, has faces that are squares; and one, the dodecahedron, has faces that are regular pentagons. In this paper Professor Artmann discusses the history of these regular polyhedra from the time when they first appear in the writings of Greek mathematicians in antiquity to their modern connections with abstract algebra.

Certainly one of the most prolific and influential mathematicians of the first part of the twentieth century was Professor G. A. Miller. This paper concentrates on Professor Miller's 44-year career at the University of Illinois. Along with some anecdotes about the Professor, is a compendium of his contributions to the founding of the American Mathematical Society and the Mathematical Association of America. This is an excellent summary of the life and work of a brilliant researcher, a dedicated teacher, and an active, responsible member of the community of scholars.

Much of the mathematical representation of electromagnetic theory began in the last half of the nineteenth century. This paper traces the development of that theory from the early work of Oersted, Ampère, Gauss, and others through the publication of Maxwell's Treatise on *Electricity and Magnetism* in 1873. It is shown that current formulations of electromagnetic theory depend on the interplay of two distinct traditions. Gauss suggested the idea of developing the entire subject from a single retarded equation for the force between moving charges which depends on the relative velocity of source and receiver. Maxwell developed the mathematical formulation of the field concepts of Faraday. These mathematical representations are now a fundamental tool of contemporary physics.

Before there was much attention being paid to multicultural education there was *Africa Counts*. This paper gives a brief description of how Professor Zaslavsky came to write one of the earliest and best known of the books now used as references for multicultural mathematics education.

Pedagogy

We certainly want our mathematics students to be literate, to be able to use language to communicate accurately and correctly. And they must have something to say. How do they attain such knowledge and skill? Professor Albree lists a number of problems with conventional term papers that he has encountered. Then he states some of his goals for students in a history of mathematics course: (1) confidence with basic library research skills, (2) organizational and writing skills, and (3) succesful completion of a major piece of writing. This paper then explains how these goals are met with a "progression of literature reports."

There is much discussion today about the use of computers in mathematics classes. This paper explains how computer laboratories and the *DERIVE* software can be used to teach the beginning calculus courses.

Teachers usually teach the way that they were taught. Lectures have been the mainstay of mathematics teaching for many years. In this paper Professor Bonnice discusses a variety of teaching techniques that can supplement or replace lectures. Included are cooperative groups, journals, and "writing to learn."

This paper discusses ways to introduce the concepts of finite mathematics and calculus in a mathematics course designed for business majors. The use of computer systems utilizing computer algebra software, graphics programs, and spreadsheets is a key element in the course design. A very nice example of building a mathematical model of a simple business decision situation illustrates most of the methodology. Professor Turner was a student of Howard Eves at the University of Maine. He concludes this paper with some recollections of Howard as a teacher.

Howard Eves—The High Elf*

Donald J. Albers

Identical Twins

Howard W. Eves

ALBERS: Where were you born?

EVES: Paterson, New Jersey, in 1911.

ALBERS: So you're kind of a big city kid.

EVES: I guess you could say so.

ALBERS: Was your early schooling in Paterson?

EVES: Yes—elementary school and early high school.

ALBERS: How about siblings?

EVES: I have a twin brother, Don, and a brother, Roland, three years older.

ALBERS: Is the twin an identical twin?

EVES: Yes.

ALBERS: Is he in mathematics, too?

EVES: No, biology. But I always feel sorry for fellows who don't have a twin brother.

---

*Professor Eves was interviewed by Donald J. Albers in his winter home in Oviedo, Florida on December 13, 1993.

ALBERS: Why is that?

EVES: It is a wonderful and unique relationship. We studied together, played together, and slept in the same room. We were inseparable and I look back upon our days together with great fondness. My twin brother is the nicest fellow I've ever known.

ALBERS: If he's identical, then you two must be very similar.

EVES: Yes. Even today we often do the same things at the same time without knowing it. How foolish we were to retire 1000 miles apart. We should have been neighbors.

ALBERS: Did your twin also end up in teaching?

EVES: He taught biology in prep schools and high schools, and was in great demand as the nature man in summer camps for boys.

ALBERS: Were the two of you ever mistaken for one another?

EVES: Oh, yes. Often. By our teachers, our parents' friends, and others. You see, all through elementary school we dressed alike. I recall an instance when even our mother failed to tell us apart. I'm sure she would have denied this, but I remember as kids she once pulled me forward to wash my face. Then she reached back to pull Don forward, but grabbed me again. I started to protest that I had just been through the ordeal, but she said, "Don't give me any trouble, I'm in a hurry now." So I went through it twice, and Don didn't get his face washed at all. But we look somewhat different now. Don parts his hair, while I just brush mine back. Also, a few years ago he badly hurt his back, so that today he is a trifle shorter than I am.

ALBERS: What sort of activities did you engage in as kids?

EVES: Oh, we played marbles, kite flying, anything to do with camping and woodcraft, featherweight back packing, canoeing, small boat sailing, and gymnastics—especially on the high bar.

ALBERS: As a young boy did you have interests in math, or was this a later development?

EVES: The first thing that interested me in math occurred, I think, in the seventh grade, when an excellent teacher introduced the class to the subject of geometrical areas. She first sketched on the board a rectangle 5 units long and 3 units wide. Then, drawing lines parallel to the sides of the rectangle through the unit divisions of those sides, she divided the rectangle into an array of small unit squares. How many of these squares are there? Well, there were 3 rows of 5 unit squares apiece, giving a total of $3 \times 5 = 15$ unit squares in all. Thus the area of the rectangle in unit squares is given by the product of its two dimensions. Clearly the same is true of any other rectangle having integral dimensions. Though at this stage, she said, it would be too difficult to prove that in any case, whether the sides are integral or not, the area of a rectangle is given by the product of its two dimensions, we will assume that this is indeed the case. Next she drew a parallelogram on the board, and dropped perpendiculars to the base from the top two opposite vertices, obtaining a rectangle easily shown to be equivalent to the given parallelogram. But the rectangle and parallelogram have the same base and altitude on that base. Therefore the area of a parallelogram is given by the product of any side of the parallelogram and the altitude on that side. Then she drew a triangle on the board, and by properly attaching one congruent to it, she obtained a parallelogram. Thus the area of a triangle is given by half the product of any side and the altitude on that side. In particular, the area of a right triangle is then half the product of its two legs. By drawing the two diagonals of a rhombus, it is seen that the area of a rhombus is given by half the product of its two diagonals. Next came the trapezoid. By drawing a diagonal of the trapezoid, thus dividing it into two triangles, we saw that the area of a trapezoid is given by the product of half the sum of its two bases and its altitude.

She then considered a regular polygon. By connecting each vertex of the polygon with the center of the polygon, the figure was divided into a number of congruent isosceles triangles, the area of each triangle being given by half the product of a side of the polygon and the apothem of the polygon. Thus the area of a regular polygon is given by half the product of its perimeter and its apothem.

Finally she drew a circle and inscribed in it a regular polygon of a large number of sides. By steadily doubling the number of sides of the inscribed polygon, we saw that the perimeter of the polygon approached the circumference of the circle, and the area of the polygon approached the area of the circle. It followed that the area of a circle was given by half the product of its circumference and its radius.

It was the neat step-by-step procedure that so fascinated me—not the individual formulas, but the inexorable progress starting from the readily assumed formula for the area of a rectangle. The other mathematics of elementary school consisted of arithmetic, and held little charm for me. Nor was I enamored by the memorization of the addition and multiplication tables. My curiosity was aroused, however, when we were taught the algorithms for long multiplication and division. It seemed remarkable that they worked, and the teacher was unable to tell us why they worked. But it was the algorithm for finding square roots that seemed most mysterious to me. The process taught us was the one where you start by marking off the digits in the given number in pairs in both directions from the decimal point. By a marvelous procedure one found the sought square root, digit by digit. Again, the teacher was unable to justify the procedure. I spent many fruitless hours trying to figure out why the procedure worked. It wasn't until I took algebra in high school and met the identity

$$(a + b)^2 = a^2 + 2ab + b^2$$

that the procedure finally fell apart. My real interest in math didn't take place until the last year in grade school, and it started with a book.

A Love Affair with Mathesis

ALBERS: Which book?

EVES: The Harpur *Euclid,* published by Longmans Green and Company. I accidentally came across it in the Paterson Public Library. It was one of those wonderful British editions of Euclid's *Elements,* and it contained only the first six books of Euclid. But the British were great at putting in challenging problems, or "riders" as they called them. I leafed through the book and found that, from a small handful of assumptions laid down at the start of the work, and to which any idiot would agree, all the rest apparently followed by pure reasoning. I was immediately reminded of my earlier experience with geometrical areas. So I checked the book out of the library, took it home and started going carefully through it. How that book engrossed me! The experience had all the aspects of a romance. It was love at first sight. I soon realized I had in my hands perhaps the most seductive book ever written. I fell head over heels in love with the goddess Mathesis.

ALBERS: Love at first sight?

Florian Cajori

EVES: Yes. And I still think that geometry is the high school student's gateway to mathematics. It's not algebra, because high school algebra is just a collection of rules and procedures to be memorized. Nothing can be called mathematics if it doesn't have the deductive feature in it, and geometry has that feature.

It's no wonder that I came to feel that for some reason Mathesis had deliberately selected me as one of her lovers, and that she must love me in return. Somehow she guided me to buy my second significant mathematics book, Cajori's *A History of Mathematics,* so that I would see the future wonderful things she could offer me. I was never jealous that she loved many others besides myself—I was just overjoyed that she loved me. And how I loved her! In time she helped me climb a great peak from which the view of the surrounding country below was breathtaking—the peak of Functions of a Complex Variable. Hand in hand we walked through a beautiful flower-strewn meadow—the field of Abstract Algebra. We bathed under a mountain waterfall—the Galois Theory of Equations. We stood side by side along a broad and beautiful flowing river—Felix Klein's Erlanger Program of Geometry. We stepped through Alice's mirror—into the land of non-Euclidean geometry. We slept in each other's arms on an ocean beach, lulled by the roar of the surf—the glistening sands of Differential Geometry. We canoed on mirror-like waters—the lake of Mathematical Logic. We walked among great redwood trees—the grove of Projective Geometry. We strolled beneath the stars on a clear and quiet night—the evening of the Philosophy of Mathematics. And we were together in many other memorable places.

ALBERS: You're getting absolutely poetic.

EVES: That's because I have been telling a love story. As the years have gone by I have aged, but Mathesis has remained as young and beautiful as ever. It was some time back that I vowed I would try to share her beauty with others, and so I have spent a large part of my life teaching in university classrooms, lyrically extolling her attractions. Perhaps I have managed to win over others to serve under the gracious goddess Mathesis.

Roots and Marbles

ALBERS: We haven't yet heard anything about your mother or your father. What do they do?

EVES: My father came over from England when he was about eight years old. His father brought him, and he had three brothers and a sister. They lived in what was considered the silk center of England. Their family made silk flags. Each family had their own loom in the attic, and they specialized in certain things. When the industry got poor there, they thought they'd come to America, to Paterson, the silk center of the United States. That's where all the silk mills were located.

And that's really the reason they all came over. My father, though, never worked in the silk industry himself. He started off as a messenger in a bank and worked his way up to vice president before he died. He died very early, in his middle 40s. He had a sudden heart attack. My mother was born in the United States of Dutch extraction.

ALBERS: Was she a conventional homemaker in those times?

EVES: My mother was one of these wonderful people who took part in her kids' games and things. She taught me to play marbles, for example, and she was a crackerjack at that.

ALBERS: Marbles!

EVES: She could, from a distance, shoot down and crack your shooter to pieces, hit it like a target. She was awfully good at it. She said the reason I wasn't so good at it was because I ate too much spaghetti.

ALBERS: You were born in 1911. Did you finish high school at the usual age of about 18?

EVES: Yes.

Virginia, Harvard, and Coolidge

ALBERS: That means you graduated from high school at the beginning of the Great Depression.

EVES: Yes. Those were very tight times. I went to Eastside High School in Paterson for three years, and the last year I attended a prep school, Chauncey Hall in Boston. My father did all kinds of bank work for clients. He invested their money and made fortunes for some of them. He made one lady into a millionaire. She had a factory that made corsets, and in appreciation for what he had done for her, she gave him quite a pile of money to send us kids to college. She did that because my father would have had difficulty doing it. So we picked our colleges. Don decided to attend the University of Virginia, so I also went there. Our older brother was already there. The three of us rented the second floor of the home of a delightful elderly retired couple, Mr. and Mrs. Long. We all graduated together in 1934. The next year my twin brother and I finally separated—he went to Columbia University for further study in biology, and I to Harvard University for further study in mathematics. Harvard was one of the very best math schools at the time. I got my master's degree at the end of my first year at Harvard. During my second year I worked as a Geometry Research Student under Julian Lowell Coolidge.

ALBERS: The famous Coolidge.

EVES: It was a great experience working under Coolidge. I also served as Assistant to Ralph Beatley. I graded and occasionally lectured for him. I also became interested in differential geometry at that time, studying the subject under the great geometer William Graustein. Most of my subsequent published papers have been in this field.

But, at that time, you couldn't get very far in differential geometry unless you were versed in tensor analysis. There was no one at Harvard in this field. In fact, it was rather rare in the United States. The best place to go, they said, would be Princeton. So I was sent to Princeton, the home of the Institute for Advanced Study. I was to go there for two years and then return to Harvard. I met Dean Eisenhart, who was the differential geometry man at the time, and several people who worked in tensor analysis. Einstein came the same year I went there, and, of course, he did a lot of work in tensors. I wasn't able to continue after the first year, because my father very suddenly died at Thanksgiving, and my financial resources came to an abrupt end.

ALBERS: So what did you do then?

EVES: I got a job as a surveyor. I became a licensed land surveyor for the State of New Jersey. I began working under someone to begin with, but later put up my own shingle. At the time there was a company called The General Housing Corporation of New Jersey. This firm would buy tracts of land and construct housing developments on them. The houses were nice, but they were inexpensive. The company wanted to make their developments look nice by putting in curved streets. That's why I got the job; other surveyors were loath to do curve work. I remember getting paid 25 cents an hour for that work, which was lucky in those days.

ALBERS: Was that about 1936?

EVES: I got my surveyor's license in 1937. From 1939 to 1941, I worked as a surveyor for the General Housing Corporation. Then I got a pretty nice job with the Tennessee Valley Authority in Chattanooga.

Back in those days all surveying computation was done by logarithms. Then some wonderful machines were developed that were about the size and heft of a standard typewriter. These were the calculators—the Marchant, the Friden, and the Monroe. The idea was to reconstruct the calculating forms from logarithmic forms to calculating machine forms. I was hired to carry out this conversion. It was a very pleasant time.

Then I got a position at Syracuse University. All the while I really wanted to teach. Oh, I did have a one-year teaching appointment at Bethany College in West Virginia, but it was just to fill in for their math man who had gone to Chicago to work on his doctorate. At Syracuse they had three math departments.

ALBERS: Three!

EVES: Yes. They had the Department of Applied Mathematics, the Department of Pure Mathematics, and the Department of Business Mathematics, each in its own building and with its own chairman and faculty. My sympathies were with the pure math department. That's where I would have preferred to be, but I was appointed to the Applied Math Department. I taught calculus and mechanics. What made my course in mechanics different from anyone else's was that I employed the axiomatic method. I started with three postulates. One, for example, was the parallelogram law of forces—the way forces combine by the parallelogram law. These postulates served for a deductive treatment of statics. Adding one more postulate, I had dynamics. This made the course in mechanics very interesting. Each of the four postulates is easily verified by a simple desk experiment performed right there in the classroom.

I've never taught mechanics since, and I kind of miss it. But I taught it for several years at Syracuse. I was classified as 1-A for the Army, and I don't know how many times I had my bag packed ready to go to Fort Dix for basic training. But the university would always say, "We need our faculty."

So I was always deferred. But I had classes of soldiers that came in and stood at attention at their desks until some corporal told them they could be seated. Then they sat down, and I taught them calculus and mechanics. They were preparing to be army engineers.

Great Moments in Mathematics

ALBERS: How long were you at Syracuse?

EVES: Three years. In 1944 I left to chair the department of mathematics at the College of Puget Sound in Tacoma, Washington. But I was there for only one year when Oregon State invited me to join its faculty. I got a promotion from Assistant to Associate Professor when I moved, and I remained there until 1951.

Then Carroll Newsom entered the picture. He was at one time Editor-in-Chief of *The American Mathematical Monthly,* and a very dear friend of mine. I think his last position was President of New York University. He had also been President of Prentice-Hall Publishing Company, and for a while he had been an educational administrator for the State of New York. As the mathematics editor of Rinehart Book Company, he invited me to write my history of mathematics for him. The first course I gave on the history of mathematics was at Oregon State, and it was called "Great Moments in Mathematics."

ALBERS: Of all things.

EVES: That's how I taught it. I dealt with 30 great moments each semester, the format being motivated by Walter Damroch's "Great Moments in Music," that he used to present on radio. Newsom suggested that I write up the course for him as a straightforward history textbook. By that time I was at Champlain College, in Plattsburgh, New York. Newsom maneuvered the move for me. Champlain College was a unit of the New York system of which he was in charge, and I became the Chairman of the Department of Mathematics at Champlain College.

ALBERS: So you moved from Oregon all the way back across the country.

EVES: Yes.

ALBERS: Because of Newsom?

EVES: Yes, largely because of his influence. Not only that, but being out West, I didn't get to see my mother, who still lived in New Jersey. I also thought I'd be able to get down to New Jersey once in a while and see my twin brother, who was then teaching in New Jersey.

ALBERS: When you were at Oregon State, did you continue to take course work? You eventually got your doctorate?

EVES: Yes, I did. Not in mathematics, but in modern physics to serve as a minor for the mathematics doctorate. I was the first doctoral candidate in mathematics from Oregon State.

ALBERS: The first one?

EVES: Yes.

ALBERS: Who actually supervised your dissertation, then?

EVES: Professor Hofstetter. I wrote two doctoral theses so they could pick whichever one they wanted. I had no trouble with that. That was the easy part for me. I'd already published a large number of papers in mathematics, pedagogical, and engineering journals. They picked one in the field of differential geometry.

ALBERS: When you say this professor supervised your thesis, was he really a differential geometer, or did you have to go back to Eisenhart?

EVES: There was somebody at Washington State University who was an expert in differential geometry and who was a personal friend of the head of our department. He refereed my work.

ALBERS: Since we're talking about these influences, let's get back to Coolidge for a minute. You said that he was a magnificent teacher for you and that he helped to increase greatly your love affair with geometry. Can you tell us a bit more about Coolidge?

EVES: When I was at Harvard, the professor I most admired was David Widder.

Now there are all kinds of methods of teaching, and some people criticize one method or another. I think it's good to have all these different methods, because somebody is often good at one method and not at another. Widder was a pure lecturer, and he was one of those professor you wouldn't think of interrupting during his lecture to ask questions any more than you would disturb Isaac Stern in the middle of a musical performance. Widder's lectures were masterfully prepared. I sat back and enjoyed his lectures enormously; they were perfect. They were given in Seaver Hall, and there was a small doorway in the middle of the front blackboard. There must have been a little office behind the blackboard. We would come to class, and at the appropriate moment Widder would emerge through the small door.

ALBERS: A dramatic entrance!

A Perfect Parabola

EVES: And he'd launch into his lecture. It was just like listening to a great violin virtuoso performing on his instrument. It was in complex variables. That's where I got my love for complex variables. If you had any questions, you asked his assistant, who, incidentally, was Herbert Robbins.

J. L. Coolidge

Now Coolidge was an entirely different person. Widder had no humor whatever, not a shred of it in his lectures. Coolidge, on the other hand, was a clown, and engaged in all sorts of antics. I remember, he had a watch on a chain hanging from his vest. He would twirl the chain around and around his index finger, and then unwind it. One day the chain broke and the watch looped across the room and smashed on a window sill. Without any hesitation he said, "Gentlemen, you have just witnessed a perfect parabola."

Which wasn't really true. A lot of people think that projectiles travel on parabolic paths. They would if the earth were a plane and the pull of gravity was constant through all heights. But when you take in the fact that the earth is spherical, and that the force of gravity varies, you get a little nose of an ellipse of which

the center of the earth is one focus, but it so closely resembles a parabola that one can't tell the difference. I didn't want to correct him, though.

ALBERS: Good idea. So he was a real showman.

EVES: Yes. I suppose when we become professors ourselves, we absorb qualities that we've admired in others. I know I have done this.

"I Wish I Never Had to Retire"

ALBERS: Well, you have quite a reputation of being more than a lecturer. In fact, I've heard some people say that for you teaching is almost a show, a performance. And that you really get into it. You really like teaching.

EVES: Oh, I wish I never had to retire. So many of my friends say they are looking forward to retiring. Not me. But when you become an octogenarian, like I am, I guess you are considered something of a health risk. Oh, I'd love to keep teaching. I miss it something dreadful.

ALBERS: What's the best part of teaching?

EVES: I think perhaps the best parts are the friendships you make with students. There's nothing quite so wonderful as the friendship that can exist between a professor and a student. I have had so many of these marvelous friendships. I could spend quite some time telling of them. You may remember Vern Hogatt.

ALBERS: Yes. I knew him well. For many years, he lived just a few miles away from me in San Jose.

EVES: I was lucky with Vern in that I was his first teacher, and we became life-long friends. I met him at the College of Puget Sound. He was a stout little fellow sitting in the front row of the class, showing rapt interest in my lectures. I found out he was just back from Germany where he'd done a stretch in the Army. We got to doing what we called our "oscillatory rambles" at night. We lived across town from one another. And he'd come and meet me at my house maybe, and we'd walk, talking mathematics with the idea of leaving off at one or the other's house, but if we weren't at the end of the discussion yet, we oscillated between houses.

I remember one time, just to see how sharp he might be, I said, "Suppose you have a two-piece screen, and you're going to cut off a corner of your room with it, and you want to cut off the biggest area behind that screen. How would you place the screen, straight across, at right angles? What are you going to do?"

I hardly finished when he stopped, and his hand came up with a finger pointing, and he said, "There's your answer." I looked down and there was a stop sign. See, it is actually a fourth of the regular octagon that yields the maximum area.

I gave a talk at the Math Club called "From Rabbits to Sunflowers." It was on the Fibonacci sequence and its esoteric properties. He became so enamored about the Fibonacci and allied sequences that he founded the *Fibonacci Quarterly.* He became the authority on those things. When I moved to Oregon, we had our last ramble together in Tacoma, and he was practically in tears that this was the end of our rambles. Well, I went to Oregon and finally found a place to live in Corvallis, went to my first classes, and there he was sitting in the first row. He said he followed me down and decided to become a student at Oregon State so we could continue our rambles.

A Snowball Down the Back!

ALBERS: There was a student that you met along the way with whom you apparently developed a very special relationship, namely, the woman seated across the room from us.

EVES: Yes. I met my wife in one of my freshman classes. She was my star student.

ALBERS: Did she go on to major in mathematics?

EVES: No, she likes biology better, and later became a Registered Nurse. She is one of those people who can be tops at almost anything.

ALBERS: How did it all come about?

EVES: I didn't know what was going on until one winter day when I was standing in the hallway talking to some of my colleagues. She passed behind me and dropped a snowball down the back of my neck. So I asked her about it afterwards, and she said, "Well, I had to get your attention some way."

ALBERS: That would do it, wouldn't it?

EVES: So we started rambling about the country-side together. And that's how it all started. I'm 23 years older than she is. At that time I had no thoughts of getting married. All I wanted to do was mathematics.

ALBERS: It looks like she changed your mind. Twenty-three years older or not, the two of you have raised five children.

EVES: One of them is at the University of Connecticut working on his Ph.D. in history; another is also at the University of Connecticut, where he just finished

his Ph.D. in philosophy. Then we have another son in Buffalo who's an attorney. And we have two girls. One is an artist who works for the University of Maine. The other daughter lives in New Hampshire, where her husband works in the fishing industry. We have five grandchildren.

MMM

ALBERS: Well, let's drop back here for a minute, if we can. You taught a history of math course for the first time at Oregon State, although your interest in history had developed earlier than that.

EVES: As you recall, the first mathematics book that I ever bought was the Harpur *Euclid,* and that book really played an enormous part in my life. The second significant math book I bought, about a year later, was Florian Cajori's *History of Mathematics.* In looking through the book, I saw all those things in mathematics that I knew nothing about. And I thought, "What a lifetime I have ahead of me learning all this material. What a wonderful field I've entered." It was with that book that my interest in the history of mathematics started. Two important things developed from that book. One was my great mathematical collection that, with affection, I have come to refer to as MMM, standing for "My Mathematical Museum." The story of this is much too long to include in this interview. The other thing that evolved from Cajori's book was the matter of The Scholar's Creed. Perhaps before we are through, I may find time to tell about it.

ALBERS: Let's hope so. When I arrived today, you showed me a couple of your watercolors.

EVES: One of my most absorbing hobbies has been watercolor painting.

ALBERS: When did you take that up?

EVES: About 1948. Prior to that I'd done nothing but pencil work, though I had always greatly admired watercolor. I actually took up watercolor on the advice of a friend—to try to cure me of being a perfectionist. Being a perfectionist is an agonizing thing, because you're never finished with anything. You always think you can make it better. When I'd write a book, I was never satisfied with what I wrote. I'd feel that I could improve my exposition, and I'd tamper with it, and never get done. But you can't do that with watercolor. There you've got to think exactly what you're going to do, do it, and never touch the painting again, otherwise you will muddy the whole painting. So I started watercolor painting, and stopped being a perfectionist.

Now I can sometimes write something and say, "That's it, and the heck with it."

ALBERS: Well, your writing, of course, is very well known in the mathematical community. Teachers love the many books you have produced over the years.

EVES: I'm flattered.

Dogs—A Superior Race

ALBERS: Well, it's true. Your *History of Math* has gone through many, many editions; your *Great Moments in Mathematics* volumes, published by the MAA, have gone on for a long time. They're MAA best sellers. Of course you've written many other books; your *Survey of Geometry* is a classic.

HOWARD EVES

Great Moments in Mathematics

Before 1650

MATHEMATICAL ASSOCIATION OF AMERICA
DOLCIANI MATHEMATICAL EXPOSITIONS—NO. 5

EVES: I've written 25 books.

ALBERS: Twenty-five! If we take into account your 80 years, that's a pretty rapid rate of production. How do you write? You say you're not as fussy or as much of a perfectionist as you were.

EVES: I first write out in pencil, then I type a rough copy, which I carefully go over before typing out the final product. All my books have been done on a typewriter, using only one finger. I'm what educators today call "a slow learner." I've tapped out about 25 books and over 150 papers in math, education, physics, and engineering. So I've done a lot of writing. But when I write a book, I try to treat it as though it's a work of art that I am producing.

But of all the books I've written, the one I regard as the most significant for math students is the one that has the poor title, *Introduction to the Foundations and Fundamental Concepts of Mathematics*. It should be called The Structural Development of Mathematics, or something like that, and be considered as a companion volume to my book on the historical development of mathematics.

ALBERS: Well, I taught from that book several years ago now, and it worked.

EVES: I dedicated the recent, third, edition to two dogs.

ALBERS: Why two dogs?

EVES: I love dogs, and I had two collies that I enjoyed so much as companions that I dedicated the book to them.

ALBERS: What's so special about dogs?

EVES: They're members of a superior race. They really are.

You can learn far more from your dog than it ever learns from you. They have marvelous characters, and are just wonderful creatures. I'll never understand why they put up with humans the way they do; we're so inferior to them. I often think of a remark made by Blaise Pascal: "The more I see of people, the better I like my dog."

ALBERS: We sort of skipped by your days at the Institute for Advanced Study; that was in the early days of the Institute.

EVES: Yes. It didn't have its own buildings then, and it was housed in Fine Hall at Princeton.

ALBERS: Were there some interesting characters roving about Fine Hall when you were there?

EVES: The Institute was my concept of Paradise. I made friends with some really remarkable people while I was there.

G.H. Hardy and Cold Tea

ALBERS: What made it so special?

EVES: The friendships I made, the colleagues I talked with, and the mathematics that was going on made it a first-rate place to be. At the top of Fine Hall there was a library with dormer nooks, and we each were assigned to one of the nooks. The year I was there, G.H. Hardy came over from England, to lecture on some Tauberian theorems. He had a nook, and every afternoon at 4:00 there would be the light tinkling of a bell. That meant there were refreshments downstairs—ice cream, coffee, tea, milk, sandwiches, etc. I was always interested in these refreshments. But we got to know that as soon as that bell tinkled, one must lean over one's desk and hold all the papers down, because like a streak Hardy would go racing by and create a vacuum behind him that would suck everything right off your desk. So one day I thought, "I'm going down early and see what that man does, and why he's so eager to get down there."

G. H. Hardy

So I kept careful track of time, and I went down five or ten minutes before the little bell was to tinkle upstairs, and I saw the lady put her finger on the bell button. She had hardly pulled her finger away when in raced Hardy. He tore over to the big silver teapot, put the back of his hand on it, and said, "Aah, stone cold."

I went over and put the back of my hand on the teapot and had to go right over to the infirmary to be treated for a burned hand. Hardy must have had asbestos hands to put them on that teapot. He claimed he couldn't get a cup of hot tea in the United States.

ALBERS: Just not hot enough.

EVES: I really think if it were any hotter it would have passed from liquid to steam. Hardy was a charming talker. He spoke such beautiful English that one was afraid to say anything in his presence for fear of sounding, by contrast, like a Chicago gangster. I had some very nice chats with him. In fact, in my mathematical museum I had one of his articles of clothing, a long white dirty scarf that in the winter he wore round his neck. I also secured a lock of Einstein's hair for my mathematical museum. I regret there is not time to tell about these acquisitions, but I have told about them in my Circle books, and will retell them in a book of recollections that I am writing.

ALBERS: Were there any other interesting people at Princeton while you were there?

EVES: Bohnenblust was there, and Church, who was mainly interested in logic. Eisenhart, of course, was there. He was also dean of the graduate school. Lefschetz was concerned at the time with topology. Albert Tucker, too. And Oswald Veblen, the great projective geometer. I got to know Veblen quite well. Veblen had a summer home in Maine, and he died there. I once visited his summer home, much in the way you would visit some shrine. Mrs. Veblen was there, and was about to entertain some lady friends by taking them for a drive along the coast. I had known Mrs. Veblen from Princeton, because she hosted biweekly teas at her home for members of the Princeton mathematics community. I asked her, "Would it be all right if I just walked around the property?" She said, "I'll leave the house door unlocked, so you can go in the house if you wish." Well, I didn't think I should do that, but I did walk around the grounds, and I came across a full-length No. 2 yellow pencil, almost brand new. And I thought, now who would carry a pencil around but a mathematician. A housewife might have one in the kitchen to write grocery lists with. This pencil, then, maybe belonged to Veblen. I put it in my mathematical museum as a doubtful acquisition.

ALBERS: When did you get to know Wiener?

EVES: I first met him in connection with a math meeting held at Yale University in 1955 or 1956. I was at that time a graduate student at Harvard. Now, in those days professors had cars and students didn't. It seems almost the other way around now. A fellow student and I heard that Wiener was driving down from M.I.T. So we went over to M.I.T., which was just down the street from Harvard, and asked if we could ride with him. He was agreeable to it, and so we met him at the appropriate time. Dirk Struik and Wiener sat in the front seat, and we two students sat in the back. It was a most hair-raising drive, for the car wove from one side of the road to the other since Wiener, while gesticulating to Struik, hardly ever had his hands on the wheel. It's a wonder that we didn't get into a wreck. Well, we got there, and it was a little early. It was raining, and we all had rain equipment. We went into this little cafeteria to get a bite to eat before the meeting started. When we finished the meal, we went to the clothes trees to get our raincoats and hats, and, of course, we students respectfully stepped back to let the professors pick up their things first. And Wiener grabbed my hat. Now, Wiener had a big head. My hat sat like a little clown hat on his head, and we all laughed. He never seemed to notice.

So I took his hat and put it on, and it came down on my nose, and I looked up at him, and I thought, now he'll see. But he never noticed a thing about it, and that's how I got his hat for my mathematical museum. He went away with my hat, and I went home with his. But when we were ready to leave, we decided we'd better make some other arrangements. We got a different ride back, and when we finally returned to Cambridge and turned on the radio the next morning, we heard that Professor Wiener's car had been stolen. He apparently forgot that he had driven to New Haven, and had gone back on the bus. When he went out to get his car to go to work in the morning, his garage was empty, so he reported to the police that somebody had stolen his car. So I had to go down there and tell him, no, it's parked down on Main Street in New Haven.

So a student and I went down to bring it back for him. He was really the typical forgetful, absent-minded type professor you read about.

Einstein's Footprints

ALBERS: Well, we've been mentioning several memorable characters here. Have we skipped over any?

EVES: Oh, I'm sure we have. I particularly remember Einstein, and could tell many anecdotes about him. I will content myself with only one.

One winter day at Princeton, after a light snowfall, I accompanied Dr. Einstein over to Fine Hall. As we were walking along, I sensed we were being followed, so I turned my head and looked back. There, about a dozen paces behind us, I saw a physics student, whom I knew, carefully putting his feet one after the other in Dr. Einstein's footprints. He did this for about half of a block. The next day I met the student and asked him why the day before he had walked in Dr. Einstein's footprints.

"Oh," he said, "I had a tough physics test coming up that morning, and I thought that if I walked in Dr.

Albert Einstein

Einstein's footprints I might perhaps catch some useful vibes." "Did it work?," I asked. "No, not at all," he mournfully replied. "Why didn't you walk in my footprints," I asked. He looked at me somewhat startled and unkindly said, "Do you think I'm that crazy?"

So it would seem that there is not much in such ideas as catching vibes, or in numerological and astrological nonsense. Consider, for example, the following facts. Garrett Birkhoff and I were fellow graduate mathematics students at Harvard. I learned that we were both born on the same day, of the same month, of the same year, in the same state of the union, and in towns with names starting and ending with like letters. Now, by the stars, we should have turned out about equally well. But Garrett became an outstanding mathematician, whereas I remained only so-so. It follows that there just can't be much in similar astrological data.

"But, ah," someone once remarked. "You didn't consider the important fact that you were born in Paterson, New Jersey, whereas Garrett was born in Princeton, New Jersey."

ALBERS: Of the huge collection of people that you've encountered, it seems that you have stories about every one of them. I understand that the people in the department here at Central Florida have taken a liking to you and that they've even given you a special name—The High Elf.

EVES: Yes.

ALBERS: What's that all about?

EVES: Well, I guess my stature brought it about. I think the man who created it was Professor Mike Taylor. When I gave talks, I used to think that H.E. stood for Howard Eves, but it stands for High Elf.

ALBERS: And you have no objections to being called the High Elf?

EVES: Oh, no, I'm flattered, really. They've been awfully nice to me at the university here. They are a wonderful bunch of people.

Problems—The Lifeblood of Mathematics

ALBERS: There's another chapter of your life that we have not yet discussed, and that's your work on problems. For several years, you served as problems editor of the *Monthly*.

EVES: I'm basically a problemist. I love problems. In fact, it's through them that mathematics lifts itself by its own bootstraps—trying to solve problems. The best thing for mathematics is to have significant unsolved problems lying· about, because in the effort to solve them, a lot of new mathematics is created. The history of mathematics shows this over and over again. Look at Fermat's Last Theorem, and all that happened before it was finally solved (we think)—such as Kummer's "ideal theory." Problems are the lifeblood of mathematics. I always try to get my students interested in problems.

ALBERS: Therefore, you've got to keep those problem studies in your history of mathematics book.

EVES: Yes, I'd be disappointed if they were removed. Some of them are difficult, I admit. But a problem isn't something that you solve immediately. Something you solve immediately is called an exercise, like solving a quadratic equation by the quadratic formula. But a true problem should be something you have to chew on, dream on for a while, and maybe sleep on it a couple of nights before you get it. Besides, the exhilaration one gets when one has finally conquered a good problem is something every mathematics student should some time experience.

ALBERS: When you yourself attack a problem, say one in algebra, do you tend first to geometrize it?

EVES: Often. That's because I usually think in terms of pictures, rather than, as most people do, in terms of words and sentences.

ALBERS: Tell us about your years as a problems editor.

EVES: For 25 years I was the editor of the Elementary Problems Department of *The American Mathematical Monthly*.

ALBERS: That's a long time.

EVES: I took it over from Coxeter. I found it extremely interesting work, for I got to know a great many mathematicians of the country and what they were currently working on. But after a while, it got to be too much for one person to handle; there were so many people sending in solutions. So I suggested that the work should be taken over by a group. That group should consist of a geometer, an algebraist, a person knowledgeable in differential equations, etc. Problems, both proposals and solutions, would then be considered by the appropriate person. The University of Maine was the first place that did this, and it has since been taken over by groups at other colleges.

ALBERS: Sharing the wealth and the work—A very good idea, I would say. Do you have any good stories about the problem business?

EVES: Many of them. For example, there is an elusive and tantalizing problem in elementary geometry known as the "butterfly problem." In an effort to obtain a number of different but elegant solutions to it, I once ran it in the Elementary Problems section of the *Monthly* (Problem E 571, May 1943). Some truly ingenious solutions were submitted, and in due time (the February 1944 issue) I published a number of them. Among the published solutions was a particularly attractive one by Professor Emory Starke, one of the country's outstanding problemists and editor, at the time, of the Advanced Problems section of the *Monthly*.

Then a few years passed by, and one day I received a letter from Professor Starke stating that that semester he had a bright Norwegian student in one of his classes who had given him a very troublesome geometry problem. Professor Starke confessed that he had spent an inordinate amount of time on the problem, but without success, and he wondered if I could perhaps help him out with it. He then stated the problem, and lo, it was the butterfly problem! For a solution I referred him, without mentioning any names, to those pages in *The American Mathematical Monthly* wherein his own elegant treatment appeared.

The Scholar's Creed

ALBERS: Let's get back to the story of The Scholar's Creed that you mentioned earlier.

EVES: There were three of us in our sophomore year at the Eastside High School of Paterson, New Jersey, who got to be known as the three math nuts. We were Al (Albert), Lou (Louis), and myself. We lived and breathed mathematics and were never happier than when we were struggling over some difficult mathematics problem.

We had heard a great deal about a remarkable branch of mathematics called "the calculus." Though today it is not uncommon to find a beginning course in calculus taught in high school, back then the practice was unheard of, and one had to await the sophomore year in college before being able to take the course.

We just couldn't wait that long, and so decided to learn the subject on our own. Now by far the most-used calculus book in those days was Granville's *Calculus.* It was the text almost universally adopted in colleges throughout the country. We accordingly ordered three copies of the famous book and, at evening meetings at one another's homes, we began teaching the subject to ourselves.

We didn't always start right off with Granville at our evening gatherings. We would sometimes discuss a new approach one of us had come up with concerning a sticky geometry problem that had been eluding us, or discuss some other mathematical matter. This procedure almost always occurred when we met at Al's home, for he usually served a round of lemonade or cocoa with cookies before we settled down.

It was at one of these sessions that I told the group of an interesting event I had read in Cajori's *History.* In reading the book I came across many highly interesting little episodes in the lives of the great creators of mathematics. One of these episodes particularly seized my attention, and it was this event that I narrated at our little session at Al's home. The episode took place in the early part of the nineteenth century at Cambridge University in England, and involved three outstanding mathematics students there, who became intimate and lifelong friends. These three students met regularly on their own to discuss mathematics, and, after some months of meeting, decided to subscribe to a vow. In great solemnity, they collectively made the following pledge:

> We vow to do our best to leave the world wiser than we found it.

And, as it turned out, the three men did precisely that.

Who were those three men? They were George Peacock, John Herschel, and Charles Babbage. Peacock was born in 1791, and was one year older than the other two. Herschel and Babbage were of the same age and both died in 1871. I won't go into the detail of just what these three men accomplished, but they certainly well lived up to their vow—they certainly left the world wiser than they found it. I was deeply impressed by their act of pledging to use their acquired knowledge in a manner that would benefit mankind, and it gave me pleasure repeating the above story to my two companions of the Paterson High School.

Al was a highly enthusiastic fellow, and when I finished the story he exclaimed, "Why don't the three of us do the same thing?" We batted the idea around a bit, but feeling we had no corner on wisdom we doubted we would be able to leave the world wiser than we found it. By further wrestling with the matter, we came up with a somewhat different pledge, which finally took the form:

> We vow never consciously to use our knowledge so as to bring harm to any of mankind, but rather to try to direct it so that the world may become a better place in which to live.

We liked this, were excited about it, and really meant it. Al produced a candle, which he lit as the sole light in the room. And, in a solemn ceremony, by candlelight, and with a triple handclasp, we made the pledge, and promised to renew it together periodically.

It happened that during the following summer I spent a week in Ohio on the farm of the grandfather of one of my rambling companions, who had no interest in mathematics. We rambled all over the farm, and spent the hot afternoons in a cool swimming hole formed by a place where the creek that ran through the farm made a sharp right-angle turn.

Boys from neighboring farms came and swam with us. We naturally got to talking about school work, and several of the fellows produced notebooks that they had. These notebooks were of all sizes and shapes, but in each of them, on a flyleaf and in beautiful Gothic lettering, appeared The Scholar's Creed by Dr. John J. Seelman. I'll never forget how impressed I was when I read the creed. Here is the main part of it, trimmed of short preliminary and short concluding paragraphs.

> I believe the knowledge I have received or may receive from teacher and book, does not belong to me; that it is committed to me only in trust; that it still belongs and always will belong to the humanity which produced it through all the generations.

> I believe I have no right to administer this trust in any manner whatsoever that may re-

sult in injury to mankind, its beneficiary, on the contrary—

I believe it is my duty to administer it singly for the good of this beneficiary, to the end that the world may become a kindlier, a happier, a better place in which to live.

Note that this creed contains the two points we made in the vow we took back in New Jersey, and that even the concluding phrase is exactly like the concluding phrase in our vow. But there is, in addition, a third point that we did not list, namely that knowledge is only on loan to us. It is claimed that some American Indians held this belief in connection with land. You do not own the land on which you live; it is only entrusted to you to use wisely and to preserve for those who come after you. In the same way, The Scholar's Creed claims that we do not own the knowledge we possess. I could scarcely wait to return to New Jersey to report The Scholar's Creed to my two friends when we should again hold one of our little gatherings.

Al and Lou were duly impressed by The Scholar's Creed. It said so much more fully and so much better what we had tried to express that we decided, on the spot, to adopt it in place of our original vow. And once again, in solemnity, by candlelight and a triple handclasp, we swore henceforth to uphold The Scholar's Creed, and to make that creed the guiding light in whatever future applications we might make of our scholastic attainments.

It has been asserted, and quite correctly I think, that our noblest ideals are formed in our youth. Can there be a teacher who has not sensed this in his or her students? A teacher will encounter young folks inflamed with strong desires in some way to better the world, whether it be toward establishing a universal brotherhood of man, advocating a wise program for conserving our planet's limited resources, or something else. However, in those cases where the teacher has been able to follow the future careers of former students, it has sadly been noticed that these earlier high ideals have often been relaxed or completely abandoned. Things happen that tempt one to lay aside one's former ideals, such as perhaps an embittering experience, or maybe some economic pressure. I dare say a teacher's parting advice to students would be, "Do not too readily compromise the fine ideals of your youth, and when you come to apply your newly acquired knowledge, recall the words of The Scholars' Creed."

Geometry – an Innate Part of Mathematics

Bruce E. Meserve
Professor Emeritus
University of Vermont

Introduction

I am particularly pleased to be able to participate in this conference celebrating the numerous professional contributions of our good friend Howard Eves. About thirty years ago Dr. Eves cited in the Preface of volume I of *A Survey of Geometry* the need for a textbook in geometry that could provide a launching ramp for students in geometry analogous to that provided twenty years earlier "by the magnificent Birkhoff-MacLane *A Survey of Modern Algebra*." He noted that as a textbook for a single course the Birkhoff-MacLane text gave undergraduate mathematics students "a glimpse into most of the important areas of modern algebra." Howard's goal of "a text that might do for geometry what the Birkhoff-MacLane text of 1941 did for algebra" is still an elusive goal. I shall not presume to identify a path to a single text or course but shall suggest some objectives for the inclusion of geometric concepts throughout the undergraduate preparation of all mathematics majors and all teachers of mathematics.

In his essay "Mathematics Education" in the September 1990 issue of *Notices of the American Mathematical Society,* William Thurston wrote:

> One feature of mathematics which requires special care is the extent to which concepts build on previous concepts. Reasoning in mathematics can be very clear and certain, and, once a principle is established, it can be relied on. This means it is possible to build conceptual structures which are at once very tall, very reliable, and extremely powerful.

> The structure is not a tree, but more like a scaffolding, with many interconnected supports. Once the scaffolding is solidly in place, it is not hard to build it higher, but it is impossible to build a layer before previous layers are in place [p. 845].

What is the place of geometry in the scaffolding for mathematics? For at least two thousand years geometry provided the standards by which the validity of other mathematical systems were measured. What aspects of geometry are essential parts of the scaffolding of the mathematics of the 21st century? What roles can and should geometry play in the study and development of mathematics in the 21st century? In exploring such questions and the development of objectives for the inclusion of geometry in the undergraduate curriculum I shall consider:

(1) the classical *roots* of geometry and the role of geometry in the classical *roots* of mathematics,

(2)   geometric *representations* of abstract mathematical systems; and
(3)   the *roles* of geometries in the discovery, development, testing, and understanding of mathematics.

I shall focus on the role of geometry as a part of mathematics rather than on past and present developments in geometry. In this context *I am thinking of mathematics as a study of patterns with the vitality of mathematics arising from the diversity of, and interrelations among, the representations, applications, and methods considered.*

Think of the general *points of view* that are used

in geometry,
in algebra and analysis,
in probability and statistics, and
in computer-based representations and methods.

While presenting a case for geometry, I am thinking of each of these four points of view as essential parts of mathematics. I use the term *point of view* to emphasize my philosophy that we should be teaching primarily *mathematics* rather than primarily a part that happens to be popular at a particular time. Forty years ago computer-based procedures were in their infancy. One hundred years ago algebra and analysis were barely reaching a mature stage. Both algebra and geometry are now of critical importance in the development of computer-based representations and methods. Assumptions that any one of these four points of view will dominate mathematics fifty years from now seems to me grossly shortsighted. Our students must be prepared to capitalize on the advantages unique to each of these four points of view.

The *Elements* of Euclid

My starting point today is a consideration of the *Elements* of Euclid not simply for its classical roots of geometry but primarily for its presentation of the classical roots of mathematics.

Part One: Plane Geometry

Book I: Lines and Line Segments on a Plane
Book II: Polygons of Equal Areas
Book III: Lines and Circles
Book IV: Circumscribed and Inscribed Figures

Part Two: Plane Geometry and Proportions

Book V: Proportions
Book VI: Plane Figures and Proportions

Part Three: Theory of Numbers

Book VII: Numbers
Book VIII: Numbers and Proportions
Book IX: Properties of Numbers

Part Four: Theory of Incommensurables

Book X: Incommensurable Magnitudes (Positive Irrational Numbers)

Part Five: Solid Geometry

       Book XI: Three-dimensional Figures
       Book XII: Three-dimensional Figures and the Method of Exhaustion
       Book XIII: Regular Polyhedrons

The descriptive titles for the thirteen books of the *Elements* are my own since Euclid did not use titles. Euclid expressed each of the 465 propositions as a rhetorical statememt. In an effort to understand the mathematical content of Euclid's *Elements* I have rephrased each of the rhetorical statements in contemporary terminology, and then restated 201 of the propositions for given drawings of geometric figures, and restated 255 of the propositions in terms of algebraic representations. Note that there are more propositions concerned with number theory or algebra than are primarily concerned with geometry. Euclid stated all propositions in terms of the geometric representations that were available to him but most of the propositions are concerned with properties of numbers or with algebraic concepts.

The 48 propositions in Book I lead up to the Pythagorean theorem and its converse. Statements regarding ratios of the areas of figures are used instead of formulas such as $A = bh$ for parallelograms and $A = (1/2)bh$ for triangles. Most of the propositions in Book II may now be viewed as geometric representations of algebraic identities, the laws of cosines, or the golden section of a line segment. Numbers and other magnitudes are represented by lengths of line segments; each product of two magnitudes by the area of a rectangle with the lengths of a pair of consecutive sides representing the quantities; each product of three magnitudes by the volume of a rectangular solid. The propositions in Book III involve properties of circles; those in Book IV, circumscribed and inscribed figures; those in Book V, magnitudes and ratios of magnitudes as in the general theory of proportions; and those in Book VI, similar triangles and constructions that we can interpret as finding positive roots of quadratic equations.

Books VII, VIII, and IX provide a foundation for the theory of numbers that is analogous to the foundation for plane geometry in Books I through VI. Book VII is concerned with divisors, multiples, prime numbers, ratios, proportions, greatest common divisors, and least common multiples. Book VIII provides an application of the theory of proportions (Book V) to the properties of numbers (Book VII). Book IX is concerned with square numbers, cube numbers, geometric sequences, proportions, the boundlessness of the set of prime numbers (Proposition IX-20), properties of odd numbers and even numbers, and an algorithm for obtaining perfect numbers.

In Book X there are 115 propositions concerned with the classification of the lengths of the edges of regular polyhedrons relative to the radius of a sphere in which they are inscribed.

Book XI is the first of three books concerned with figures in three-dimensional space. The propositions in Book XI involve parallelism and perpendicularity of lines and planes, the congruence of opposite faces of a parallelepiped, and the ratios of volumes of parallelepipeds and related polyhedrons. The propositions in Book XII are concerned with the development of properties of polyhedrons and the use of the method of exhaustion, in the sense of using up all parts of a figure, to extend those properties to cones, cylinders, and spheres. The propositions in Book XIII, the final book, are concerned with the construction of the five regular polyhedrons (the tetrahedron, cube, octahedron, dodecahedron, and icosahedron) and the observation that there can be no other regular polyhedrons. Whether Euclid's use of constructions in the development of regular polyhedrons will provide useful insights for the development of regular figures in higher-dimensional spaces remains to be seen.

The careful structure of the *Elements* is indicated by the dependence, directly or indirectly, of the last proposition, Proposition 18, upon 16 of the previous 17 propositions of Book XIII, 33 of the 48 propositions in Book I, and many of the propositions in the intervening books.

With this quick survey of the scope of Euclid's *Elements*, let us shift our attention to geometric representations and consider first geometric representations of numbers.

Geometric Representations of Numbers

In Books II, V, VII, VIII, IX, and X Euclid demonstrated that the properties of numbers and magnitudes that he needed in the development of mathematics in terms of geometric representations were available in the axiomatic system that he had established. For example, the commutative property of multiplication was established for positive integers in Proposition VII-16:

> If each of two numbers is multiplied by the other and the products are numbers, then the products are equal.

The condition that the products of the numbers be considered as numbers reflects the representation of such products either by areas of rectangles with line segments that represent the given numbers as sides or as fourth proportionals $1 : a = b : ab$ that could be represented by lengths of line segments.

Euclid viewed ratios as *relations between numbers* rather than as numbers themselves. For example, consider Proposition VI-17:

> If each of two given numbers is multiplied by the same number and the products are numbers, then the ratio of the products is the same as the ratio of the corresponding given numbers.

In other words, if a, b, and m are positive integers, then $ma : mb = a : b$.

Euclid's use of lengths of line segments as magnitudes in Book II and his proofs in Book V that magnitudes had many of the properties of numbers were early forerunners of our concept of *positive real numbers*. In Book X Euclid used straightedge and compass constructions of line segments to introduce the irrational magnitudes that he needed to construct regular polyhedrons. Since the line segments are constructible, their lengths can be expressed in terms of positive integers, rational operations and radicals for square roots.

Two magnitudes are defined to be *commensurable magnitudes* if their ratio is equal to a ratio of positive integers. Magnitudes that are not commensurable are *incommensurable magnitudes*. Any two incommensurable line segments such that the areas of the squares on the line segments are commensurable, are called *square rational line segments*. For example, the areas, 9 and 2, of the squares on line segments of lengths 3 and $\sqrt{2}$, are commensurable and the line segments are square rational line segments. For any two incommensurable square rational line segments Euclid constructed:

the *medial line segment* $\sqrt{3\sqrt{2}}$, that is, the mean proportional,

the *binomial line segment* $3 + \sqrt{2}$, that is, the sum, and

the *apotome line segment* $3 - \sqrt{2}$, that is, the positive difference.

In general, Euclid constructed line segments to represent seven distinct and mutually exclusive classes of irrational magnitudes that can be represented algebraically as follows.

For r^2, s, t, and u positive rational numbers such that $\sqrt{s}$ and $\sqrt{u}$ are incommensurable irrational numbers:

medial line segments (X - 21): $r\sqrt{\sqrt{s}}$

binomial line segments (X-36): $r + rt\sqrt{s}$

first bimedial line segments (X-37): $r\sqrt{\sqrt{s}} + t / \left(r\sqrt{\sqrt{s}} \right)$

second bimedial line segments (X-38): $r\sqrt{\sqrt{s}} + t\sqrt{u\sqrt{s}}$

For b^2, s, and t positive rational numbers such that $\sqrt{1 + s^2}$ and $\sqrt{t}$ are irrational numbers:

major irrational line segments (X-39):

$$\sqrt{\frac{b^2}{2}\left(1 + \frac{s}{\sqrt{1+s^2}}\right)} + \sqrt{\frac{b^2}{2}\left(1 - \frac{s}{\sqrt{1+s^2}}\right)}$$

consequents of a rational area and a medial area (X-40):

$$b\sqrt{\frac{1}{2\sqrt{1+s^2}} + \frac{s}{2\left(1+s^2\right)}} + b\sqrt{\frac{1}{2\sqrt{1+s^2}} - \frac{s}{2\left(1+s^2\right)}}$$

consequents of two medial areas (X-41):

$$b\sqrt{\frac{\sqrt{t}}{2}\left(1 + \frac{s}{\sqrt{1+s^2}}\right)} + b\sqrt{\frac{\sqrt{t}}{2}\left(1 - \frac{s}{\sqrt{1+s^2}}\right)}$$

or $b\sqrt{(1/2)\left(\sqrt{s} + \sqrt{s-t}\right)} + b\sqrt{(1/2)\left(\sqrt{s} - \sqrt{s-t}\right)}$

where $\sqrt{s}$ and $\sqrt{s-t}$ are incommensurable irrational numbers.

For each of the six sums Euclid considered the positive differences and, using only geometric representations, proved that the thirteen classes of irrational magnitudes were mutually exclusive. He also identified, relative to a given square rational line segment, six classes of binomial line segments and six classes of apotome line segments. As mentioned earlier all of this was done to obtain classifications of the lengths of edges of regular polyhedrons relative to the lengths of the radii of their circumscribed spheres.

Proposition X-112 is equivalent to a statement that if a rectangle with a rational area has a line segment of length $r + rt\sqrt{s}$ as one side, then the other side is a line segment of length $(kr)-(kr)t\sqrt{s}$ for some positive rational number k. Note the similarity of these concepts with the modern use of rationalizing factors as indicated, for instance, by

$$\left(a + \sqrt{b}\right)\left(a - \sqrt{b}\right) = a^2 - b \text{ and } \left(\sqrt{a} + \sqrt{b}\right)\left(\sqrt{a} - \sqrt{b}\right) = a - b.$$

Even though decimal representations for real numbers have now replaced Euclid's geometric representations, let us not dismiss Euclid's very impressive work lightly. It is a prelude to an active role of numbers in the development of geometry and an active role of geometry in the development of number systems. The introduction of coordinates in geometry by Descartes and Fermat early in the 17th century led to a major revolution in mathematics. We now recognize that operations can be defined so that the geometry of points on a Euclidean coordinate line and the algebra of real numbers are isomorphic.

Geometric Representations of Algebraic Concepts

At the time of Euclid all algebraic statements were expressed in words, that is, the algebra of that time was a *rhetorical algebra* rather than a *symbolic algebra*. The geometric representations of algebraic identities in Book II are sometimes described as *geometric algebra*. In our modern age of digital representations and computer processing of the masses of data obtained it is important that we remind ourselves that at the time of Euclid all numbers were positive integers and that algebraic symbolism had not yet evolved. Thus we need to set aside our reliance on manipulations of algebraic symbols as we consider Euclid's development of mathematics. Euclid relied solely on his postulates, common notions, definitions, and a few unstated, that is, tacit, assumptions.

The Pythagorean theorem provides a prelude for distance relations $ds^2 = dx^2 + dy^2$ and for generalized distance relations in higher dimensions and in other geometries.

Euclid's ratios were relations that served as forerunners of our concepts of fractions. For example, Proposition V-24:

If $a, b, c, d, e,$ and f are given such that $a : b = c : d$ and $e : b = f : d$, then $(a + e) : b = (c + f) : d$.

corresponds to our addition of fractions.

From a modern point of view it is tempting to think of the propositions in Book V as trivial applications of the laws of algebra. However, the proper place of these propositions in the *development of mathematics* is as geometric foundations for many of the laws of algebra.

The constructions in Book VI of line segments of length x that represent a third proportional, a fourth proportional, and a mean proportional,

$$a : b = b : x \quad a : b = c : x, \quad a : x = x : b$$

are equivalent to solutions of equations of the form

$$ax = b^2 \qquad ax = bc \qquad ab = x^2$$

in terms of geometric representations. Instead of asserting that an equation of the form $x - a = b$ can be solved, Euclid would have used line segments to represent magnitudes and written

a line segment can be constructed that exceeds a given line segment by another given line segment.

Instead of asserting that any quadratic equation of the form $x^2 + px = q^2$ can be solved, Euclid reasoned, in Proposition VI-29, in terms of the area of a rectangle determined by line segments of lengths $x + p_2$ and x being equal to the area of a square on a line segment of length q, that is, $x(x + p) = q^2$. If a relation that we represent by a quadratic equation with real coefficients has a positive root, Euclid provides a construction of a line segment whose length represents a positive root of the given relation. His approach is remarkably complete and entirely geometric. For example, his Proposition VI-27 is equivalent to establishing the necessity of a nonnegative discriminant if there is to be a real root.

Euclid and the early Greeks did not directly "solve quadratic equations" because they did not have the symbolic algebra for writing quadratic equations. However, they did solve *for a positive root* rhetorical problems equivalent to the problems that we solve by finding positive roots of quadratic equations with real coefficients. Quadratic mathematical relations have not changed - only our representations of the relations have changed. Certainly the quadratic formula is easier than the construction in Proposition VI-28 for finding two numbers whose sum and product are given positive numbers. However, we should not lose sight of the role of geometry in providing a tested basis for later algebraic rules and procedures.

The preoccupation of early algebraists with homogeneous equations reflected their representation of each product of two numbers by an area and each product of three numbers by a volume. Then each equation was

a statement about lengths of line segments, that is, a linear equation, or
a statement about areas, that is, a quadratic equation, or
a statement about volumes, that is, a cubic equation.

The geometric representations for numbers and magnitudes in the *Elements* and the establishment of the basic properties of numbers, magnitudes, ratios of numbers, and ratios of magnitudes set the stage for the development of symbolic algebra. In the 17th century as Descartes and Fermat were introducing coordinate geometry, and as magnitudes such as x^2 and x^3 were slowly gaining recognition as having the same properties as numbers, Rene´ Descartes (1596-1650) described the close relationship of algebra and geometry as follows:

algebra is needed to avoid a dependence on the appearance of geometric figures, and
geometry is needed to give meaning to the operations of algebra.

In his article "What is Geometry?" in the October 1990 issue of *The American Mathematical Monthly* S.S. Chern writes

"Fermat went on to introduce some of the fundamental concepts of the calculus, such as the tangent line and maxima and minima."

"From two dimensions one goes to n dimensions, and to an infinite number of dimensions. In these spaces one studies loci defined by arbitrary systems of equations. Thus a great vista was opened, and geometry and algebra became inseparable."

In view of the danger of our present students and colleagues losing sight of the place of geometry as a main pillar in the scaffolding of mathematics, I interpret Dr. Chern's remark as meaning that although algebra and geometry appear distinguishable in the nature of their representations, they are inseparable in their roles in the ongoing development of mathematics.

Geometries as Models of the Physical Universe

The physical universe appears to most people to be a three-dimensional universe. From the time that they first recognize the existence of physical objects, all animate creatures appear to classify objects as edible or not edible, friendly or unfriendly, and so forth. In their efforts to understand the physical universe, and in most cases to exploit it for their own purposes, human beings have classified physical objects by every conceivable criterion. Classifications as to number, position, size, shape, and changes in these attributes seem to me to provide primitive bases for mathematics. From these very crude beginnings mathematics has evolved and is continuing to evolve through processes of

modeling, generalization, and abstraction to extend the domain of mathematics;
axiomatization to provide a logical foundation for mathematics; and
computational procedures that are increasingly useful and precise.

In this context I am thinking of the arithmetization of analysis a century ago as a form of modeling.

The numerical approach of the Babylonians was followed by the geometric approach of the classical Greeks. Note that a first level of abstraction arises, for example, by the introduction of

points for the positions of objects, or locations, in space;
surfaces of objects as sets of points;
open curves and closed curves as paths of objects;
lines and circles as special classes of curves;
planes as special classes of surfaces; and
numerals and lengths of line segments as numbers.

One of the roles of geometry was to show that lengths of line segments had many of the same properties whether they represented multiples of a given unit segment, ratios of those multiples, or magnitudes. The concept of geometric figures as abstractions from objects in the physical universe provides the basis for the consideration of a geometry as a model of the physical universe.

The use of a drawing of a geometric figure to represent an arbitrary point on a line segment or any positive real number provided a basis for the generalization of the natural numbers and the development of algebra. In this sense, whether by way of arithmetic or geometry, algebra represents a second level of abstraction that is twice removed from the realities of the physical universe. This status of algebra as a second level of abstraction has little, if any, significance to mathematicians but for students it means that extensive prealgebra experiences, both numerical and geometric, are needed before algebra can be fully understood. Note also that it is the status of calculus as a third level of abstraction that causes many students difficulty. Most such difficulties can be avoided by making sure that all students are able to interpret the concepts of calculus in terms of both geometric and algebraic representations. The use of arithmetic, geometric, and algebraic models is undoubtedly clear to everyone. Although axioms seem most often mentioned in conjunction with geometry, many established areas of mathematics have their own axiomatic foundations - often based on those for a geometry or an algebra. Euclid used his axiomatic system to develop not only plane geometry but also solid geometry and the prerequisite algebraic properties.

Over the last 2300 years several weaknesses have been found in Euclid's axiomatic system. For example, we now know that not all terms can be defined, that is, there must be some primitive terms such as point, line, incidence, and betweenness, that are accepted without definition and have the properties imposed on them by the axioms. Relative to modern standards Euclid's use of constructions did not properly specify existence, order relations, and the continuity of lines and circles. All such weaknesses now appear to have been taken care of with the identification of primitive terms, the addition of new axioms, and revisions of some of the proofs.

For two thousand years there was an extensive search for a proof of Euclid's parallel postulate. Early in the 18th century Girolamo Saccheri studied the properties of the vertex angles of an isosceles birectangular quadrilateral where the angles at A and B are right angles and the sides AC and BD are congruent.

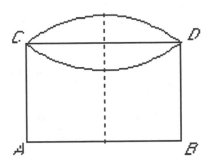

Saccheri was able to prove from Euclid's other postulates that the angles at C and D must be congruent and that the line joining the midpoints of the sides AB and CD is perpendicular to both of these sides. However, as shown in the figure, there remained three possibilities for the side CD according as the angles at C and D were acute angles, right angles, or obtuse angles. Saccheri finally convinced himself that only the right angle hypothesis could hold and published his results in 1733 in a little book bravely entitled, "Euclid Freed of Every Flaw."

One hundred years later the work of Janos Bolyai and Nicolai Lobachevsky made the existence of the geometry associated with Saccheri's acute-angle hypothesis, that is, hyperbolic geometry, logically unavoidable. The existence of more than one geometry created a major revolution in the development of mathematics. Philosophers were forced to recognize that there could be more than one axiomatic system. It was no longer necessary to study only representations of the physical universe. Mathematicians were freed to explore the development of different mathematical systems -- different geometries, different algebras, and, in general, different abstract axiomatic systems.

The axioms of any axiomatic system are considered to be consistent if there exists a physical representation for them. In this context as a model of the physical universe Euclidean geometry often serves as the testing ground, sort of a mathematical proving ground, for other axiomatic systems.

The early Pythagoreans considered geometry to be the study of figures at rest and astronomy to be the study of figures in motion. Modern astronomers are continuing to use geometries in their explorations of the nature of the universe. They even have theories as to the future of the universe depending on whether its underlying geometry is Euclidean, hyperbolic, or elliptic. Although the most widely accepted theory involves hyperbolic geometry, Euclidean geometry, because of its relative simplicity, continues to be the preferred geometry for the study of relatively small regions of the universe.

The Roles of Geometry

We have noted the classical role of geometry in the *Elements* and the basic role of geometric representations in the preliminary development of number theory and algebra. Before we consider some of the other roles of geometry in the discovery, development, testing, and understanding of mathematics let us take a quick view of contemporary geometry as seen by an internationally recognized geometer. I cite the conclusion of Dr. Chern's article in the October 1990 issue of the *Monthly*:

"Contemporary geometry is thus a far cry from Euclid. To summarize I would like to consider the following major developments in the history of geometry:

1) Axioms ([based on the work of] Euclid)
2) Coordinates ([based on the work of] Descartes and Fermat)
3) Calculus ([based on the work of] Newton and Leibniz)
4) Groups ([based on the work of] Klein and Lie)
5) Manifolds ([based on the work of] Riemann)
6) Fiber bundles ([based on the work of] Eli Cartan and Whitney)"

"A property is geometric if it does not deal directly with numbers or if it happens on a manifold, where the coordinates themselves have no meaning. Going on to several variables, algebra and analysis have a tendency to be involved in geometry."

"...Recent developments in theoretical physics, such as geometric quantum theory, string theory, etc. are pushing for a much more general definition of geometry..."

"It is satisfying to note that so far almost all of the sophisticated notions introduced in geometry have been found useful."

In the language of William Thurston Dr. Chern describes geometry as a conceptual structure "which is at once very tall, very reliable, and extremely powerful." However, my goal is not to describe the achievements of geometry but rather to provide a rationale for the inclusion of concepts of geometry throughout the undergraduate curriculum. My general approach is:

to illustrate some of the ways in which geometry has served in the scaffolding of mathematics, while discussing these examples to identify the corresponding roles of geometry, and
to conclude with a suggested list of objectives for the inclusion of concepts of geometry throughout the undergraduate curriculum

My selections of examples of ways in which geometry has served in the scaffolding of mathematics represent only a miniscule sample of the many examples available. I urge each of you to add examples from your own experiences.

We have already noted the roles of geometric representations and theorems of geometry as aids to our *development* and *understanding* of:

the distributive law for multiplication over addition,
the commutative law for multiplication,
the expression for the square of a binomial sum,
the law of cosines,
the determination of roots of quadratic statements of equality,
relations among ratios of numbers,
relations among magnitudes and their acceptance as positive real numbers,
properties of greatest common divisors and least common multiples,
properties of prime numbers and composite numbers,
the identification of classes of irrational numbers, and
the recognition of the square of a number as a number and the cube of a number as a number.

In the 17th century, analysis was understood by Newton and Leibniz to be concerned with continuous magnitudes such as lengths, areas, and volumes. The basic representations of these magnitudes were geometric. The study of tangent lines and areas under curves led to the development of differentiation and integration. The usefulness of a strongly intuitive and geometric background is exemplified by the outstanding contributions of G. F. B. Riemann (1826-1866), J. W. R. Dedekind (1831-1916), and Georg Cantor (1845-1918).

In 1854 Riemann urged a global view of geometry as a study of manifolds of any dimension and in any kind of space. Points were represented by n-tuples of numbers that were combined according to specified rules. One of the most important rules was the procedure for finding the distance between any two infinitesimally close points. In Euclidean three-dimensional geometry

$$ds^2 = dx^2 + dy^2 + dz^2$$

but any homogeneous quadratic expression in the differentials of the coordinates with constant coefficients could be used as a metric in a Riemannian space. The study by Riemann and others of curvature in a Riemannian space, or manifold, ultimately made the theory of general relativity possible. Riemann also used a geometric approach in refining the definition of integrals and for developing Riemann surfaces, which provide an ingenious scheme for making functions one-to-one rather than multivalued.

Dedekind, in his search for the meanings of limits of convergent infinite series of rational numbers that did not converge to rational numbers, initially concluded that the usual guidelines from geometry should be avoided in order to be rigorous. Then he realized that the distinguishing property of continuous linear geometric magnitudes was not a vague hang-togetherness but rather the division of the line segment into two parts by a single point of the segment. He reasoned that if the points of any line segment are separated into classes such that

       every point of the segment belongs to exactly one class, and
       every point of one class is on the left of every point of the other class,
       then there is one and only one point that brings about this partition of the line segment.

This geometric insight and the assumption that the points of a line can be placed in one-to-one correspondence with the real numbers enabled Dedekind to extend the domain of rational numbers to form the continuum of real numbers. This approach was published in 1872.

Throughout most of the 19th century the one-to-one correspondence of the set of natural numbers and the set of squares of natural numbers, an obviously "small part" of the set of natural numbers, was considered to be a paradox. Dedekind was the first to recognize, again published in 1872, that the equivalence of a set with one of its proper subsets was a characteristic property of infinite sets, and that this property could be used to define infinite sets. Two years later, in 1874, Cantor was studying transcendental numbers and used a geometric representation to show that there are as many real numbers on any open interval as there are real numbers.

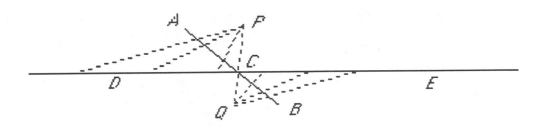

Note that for an open line segment AB that intersects a line DE at a point C Cantor used as centers of projection a point P such that $AP \parallel DE$ and a point Q on PC such that $QB \parallel DE$.

These examples of the contributions of Riemann, Dedekind, and Cantor are suggested as illustrations of the role of geometry in the *discovery* of mathematics.

Euclid's geometric development of properties of ratios provided tested procedures for working with positive rational numbers represented by fractions. In the 19th century Dedekind cuts and Cantor's nested intervals provided the insights needed for understanding the place of irrational numbers in the set of real numbers. The Cantor-Dedekind axiom of the one-to-one correspondence of the points on a Euclidean line and the real numbers has provided a basis for improving our understanding of both Euclidean lines and real numbers. The opportunity to use either geometric or algebraic representations has an obvious value in both research and the teaching of mathematics, since some people favor the left hemispheres of their brains and others favor the right hemispheres. Everyone has opportunities, and should be encouraged, to develop both hemispheres of their brains – that is, to make use of all of their mental faculties.

It seems to me that a major problem arises when we, as teachers and authors, deprive students of opportunities to recognize the distinct nature and relative advantages of both geometric and algebraic representations. In order for students to have an opportunity to profit from the use of mathematical models we must emphasize the use of both geometric and algebraic models. However, it seems that many teachers and writers have reverted to the Pythagorean "All is number" philosophy. Let me give you two specific examples based upon statements that seem to me to reflect a somewhat common cultural bias.

Richard Trudeau has written a very informative and thought-provoking book "The Non-Euclidean Revolution" published in 1987 by Birkhauser. I highly recommend the book to you even though I take exception to two of his statements. On page 245 he writes,

"The graph of $3x + 4y = 6$ is not really a Euclidean straight line. At root it is a set of pairs of real numbers, specifically the set of all pairs (x, y) of real numbers for which $3x + 4y$ is equal to 6 ..."

After a paragraph of complimenting the readers who have been fortunate enough to have teachers with this insight, he continues his point that Real Analysis is the ultimate proving ground for tests of consistency of axiomatic systems. I'll return to that point but first let's think about the insistence that a graph is a set of pairs of real numbers.

Mathematics has thrived on precise definitions. To capitalize on that strength I suggest that

(1)   in a two-dimensional Euclidean geometry model the *graph* of $3x + 4y = 6$ is a straight line; *to graph* the linear equation is to draw that line on a coordinate plane;

(2)   in a two-variable real analysis model the *solution set* of $3x + 4y = 6$ is a set of ordered pairs of real numbers; *to solve* the linear equation is to identify that set of ordered pairs by examples and in terms of a single real parameter.

Relative to the consistency of axiomatic systems Trudeau summarizes the present status of Euclidean geometry on page 244

"Logical weaknesses were found in the *Elements* itself; though these were successfully shored up with extra axioms and new demonstrations..., the book's reputation for logical rigor was permanently tarnished. The possibility that Euclidean geometry might be inconsistent was unthinkable no longer. An alternative to the traditional model had to be found. The natural choice was "analytical geometry" a part of Real Analysis (*not* of geometry)."

My concern is for his argument that Real Analysis is the natural choice. After establishing the algebraic interpretations of points, lines, and circles, he writes on pages 245 and 246:

> "Since the model turns any contradiction in Euclidean geometry into a contradiction in Real Analysis, Euclidean geometry is consistent if Real Analysis is."

> "There the matter rests. As I write the consistency of Real Analysis is still an open question, and is likely to remain so."

I see no objection to any writer identifying the altar at which he kneels. My point is that, as Trudeau has written but does not seem willing to admit, Euclidean geometry and real analysis serve equally well as testing grounds for consistency. The choice between them can be made as a matter of convenience or preference. But, as illustrated by the works of Riemann, Dedekind, and Cantor, the elimination of geometry as a viable alternative can only handicap future students and mathematicians.

Let me reiterate that I am not "shooting at" Trudeau. He has written a fine book. Rather I am using his statements of a common contemporary point of view, with which I strongly disagree, to argue for a broader point of view.

Algebraic representations have dramatically simplified many aspects of geometry. Algebraic structures such as modules and groups have been very useful in recognizing the structures of, and relations among, different geometries. Both algebraic representations and algebraic structures should be used extensively in the study of geometry. Similarly, geometric representations should be used extensively in the study of algebra. Indeed, geometric concepts should be widely used in the study of most areas of mathematics. For example, consider the use of a hypercube in 52 dimensions in the 1990 proof by Persi Diaconis and David Bayer that seven shuffles are needed and are sufficient for a thorough mixing of a deck of 52 cards.

I do not argue that geometry has a major role in every aspect of mathematics. Rather I suggest that

1. geometry has had a major role in the development of mathematics,
2. geometry continues to have a major place in the scaffolding of mathematics, and
3. all mathematics majors need to understand geometry as an innate part of mathematics.

Objectives for the Inclusion of Geometric Concepts in the Undergraduate Curriculum

In conclusion, I'll propose several objectives for the inclusion of geometric concepts in the undergraduate curriculum of all mathematics majors and all teachers of mathematics. In that sense I'll suggest a pathway to Dr. Eves' elusive goal thirty years ago.

1. Relative to the major developments in the history of geometry:

Axioms	Coordinates	Calculus
Groups	Manifolds	Fiber bundles

that were cited by Dr. Chern each student should

(a) be able to identify the axioms and primitive terms, to define needed terms, and to prove suggested theorems in at least one synthetic geometry,
(b) be able to identify the axioms and primitive terms, to define needed terms, and to prove suggested theorems in at least one analytic geometry,

    (c)   be able to provide examples of the use of geometry in calculus and examples of the use of calculus in geometry,

    (d)   be able to provide examples of the use of geometry in algebra and examples of the use of groups in geometry, and

    (e)   at least for those students planning to do graduate work in mathematics, be able to give examples of the role of manifolds in the development of mathematics and be able to give examples of the role of fiber bundles in the development of mathematics.

2.   To gain an understanding of the existence of a variety of geometries and the usefulness of geometric representations as a basis for conjectures each student should be able to use representations in Euclidean three-space for Euclidean, spherical, elliptic, and hyperbolic two-dimensional geometries to compare and illustrate properties of each geometry relative to intersections of lines, parallelism of lines, angle-sums of triangles, and absolute units of angle measure, distance, and area.

3.   To recognize geometry as a part of mathematics each student should be able

    (a)   to make effective use of geometric representations in each area of mathematics that the student studies,

    (b)   to describe the role of geometry in the development of each area of mathematics with which the student has had experience, and

    (c)   to conjecture roles for a geometric point of view in each area of mathematical research in which the student has had experience.

4.   To recognize the place of Euclidean plane geometry in the study of geometry each student should

    (a)   be able to extend analytic Euclidean plane geometry to, and prove theorems in, Euclidean geometries in three-, four-, and n- dimensions, and

    (b)   be able to generalize Euclidean geometry, or specialize projective geometry, to show the relations among projective, affine, equiareal, similarity, and Euclidean geometries.

5.   To obtain background experience for working in an axiomatic system each student should understand and be able to use

    (a)   the relations among the truth values of a conditional statement, its converse, its inverse, and its contrapositive, and

    (b)   the common rules for logical deductions and proofs.

6.   Each student, especially prospective teachers of secondary school mathematics, should be able

    (a)   to explain an isomorphism of the set of real numbers and the set of points on a Euclidean line,

    (b)   to use synthetic, analytic, and all other available methods, such as uses of counterexamples, transformations, vectors, and calculus, in problem solving and in proofs, and

    (c)   to make and test conjectures.

In order to achieve such objectives mathematicians who are actively using geometry, especially those using manifolds and fiber bundles, must convey the nature of their work to the mathematical community. It may not be necessary for these mathematicians to write textbooks, but their insights are essential in order to convey a proper perspective. Many of us have tried to develop a single course in geometry but we have failed to present geometry in a broad perspective and we have failed to convince our colleagues that our proposed course is needed by most mathematics majors.

It seems to me that our basic goal should be the inclusion of geometric concepts throughout the undergraduate curriculum. Not only would it be much more feasible to develop a course in geometry for students who had made extensive use of geometry in their study of calculus and linear algebra but, as we are learning from the use of computer graphics, these students would also have a much bettter understanding of calculus and linear algebra.

I believe that an understanding of geometry is essential if our students are to work creatively in their selected areas of specialization. At an intuitive level I am arguing that both pictorial and symbolic representations are innate parts of mathematics. My goal today has been to establish the necessity for acquiring a background in geometry and to suggest objectives for the inclusion of geometry throughout the undergraduate curriculum of all mathematics majors and all teachers of mathematics. Whatever structure is used, the treatment needs to be sufficiently broad in its scope and emphasis to provide students with an overall concept of geometry as an innate part of mathematics.

Connecting Geometry and Number Theory

Peter Hilton
SUNY Binghamton
Binghamton, New York 13902-6000

Jean Pedersen
Santa Clara University
Santa Clara, California 95053

Dedicated to Howard Eves on the occasion of his 80th birthday

INTRODUCTION

We begin with a few words as to why we have chosen this topic, and this presentation, as our contribution to a conference held in honor of Howard Eves.

First, the topic itself, while essentially elementary (that is, requiring only the mathematics encountered in undergraduate courses), exhibits the features most characteristic of worthwhile mathematics. There is a clearly defined objective, but the attempt to achieve that objective leads us into unexpected areas of mathematics apparently remote from the original context. Moreover, we advance in those areas in directions *not* dictated by the original objectives but by the internal dynamic proper to those areas. On the other hand, we also advance in response to stimuli and formulations provided by our geometrical problem.

Second, we have tried to exhibit a pedagogical strategy which makes plain not only how we selected the questions asked but also how we selected the methods and arguments required to obtain answers. We do not believe it effective to discuss mathematical pedagogy in the abstract – certainly we do not know how, effectively, to do so.

Third, by really engaging our students in the mathematical activity involved, including, certainly, the actual construction of the regular polygons and the actual calculation of quasi-orders (see Section 4), we can make them see that mathematics is enjoyable – it is both intriguing and important, both esthetically pleasing and immensely powerful.

Fourth, we feel that the historical setting of the problem is in keeping with much of Professor Eves's own work. Let us start with the history.

Our story begins with the Greeks and their fascination with the challenge of constructing regular convex polygons – that is, those polygons with all sides of the same length and all angles equal. They wanted to create these polygons using only an unmarked straight edge and compass. As is well known, drawing an *exact* geometric figure with these restrictions is called a *Euclidean construction* – and the unmarked straight edge and compass are called *Euclidean tools*. The Greeks (ca. 350 B. C.) were successful in devising Euclidean constructions for regular convex polygons having N sides,[1] where N is a power of 2 or

$$N = 2^c N_0 \,, \quad \text{with} \quad N_0 = 3, 5 \text{ or } 15 \text{ and } c \geq 0.$$

(Of course, we need $N \geq 3$ for the polygon to exist.)

No further progress appears to have been made in the next 2000 years, until Gauss (1777-1855) proved that a Euclidean construction of a regular N-gon is possible *if and only if* the number of sides N is of the form $N = 2^c \prod \rho_i$, where $c \geq 0$ and the ρ_i are distinct Fermat primes – that is, primes of the form $F_n = 2^{2^n} + 1$.

Gauss's discovery was remarkable – and, of course, it tells us precisely which regular N-gons admit a Euclidean construction. However, it turns out that not all Fermat numbers are prime. Euler (1707 - 1783) showed that $F_5 = 2^{2^5} + 1$ is not prime and, in fact, to this day the only known prime Fermat numbers are

$$F_0 = 3, \quad F_1 = 5, \quad F_2 = 17, \quad F_3 = 257 \text{ and } F_4 = 65537.$$

Thus, even with Gauss's contribution, there exists a Euclidean construction of a regular N-gon for very few values of N, and even for these N we do not in all cases know the explicit constructions.

Despite our knowledge of Gauss's work we still would like to be able to construct (somehow) *all* regular N-gons. Our approach is to redefine the problem raised by the Greeks so that, instead of exact constructions, we will be concerned with arbitrarily good *approximations.* Our aim is to show how one can, simply and systematically, construct an *approximation* (to any degree of accuracy) to a regular convex N-gon for *any* value of N.[2] In fact, we give explicit (and uncomplicated) instructions involving only the folding of a straight strip of paper (like adding machine tape), in a prescribed and repetitive manner.

---

[1] We will refer to such polygons as regular N-gons, and we may even suppress the word 'regular' if no confusion would result.

[2] In this respect we will argue that in practice the approximations we obtain by folding paper are quite as accurate as the real world constructions with a straight edge and compass – for the latter are only perfect in the mind. In either case the real world result is a function of human skill, but our procedure, unlike the Euclidean procedure, tends to reduce the effects of human error.

Although the construction of regular convex N-gons would be a perfectly legitimate aim by itself, the mathematics we encounter is generous and we achieve much more. In the process of making what we call the *primary crease lines* on the tape used to construct regular N-gons we naturally obtain tape which can be used to fold certain (but not all) regular star polygons. However, it is not difficult to add *secondary crease lines* to the original tape in order to obtain tape that may be used to construct the remaining regular star polygons. And there is even more! For the mathematics we encounter in the process of constructing these regular polygons leads quickly and naturally to ideas in elementary analysis; and the question of which convex or star polygons we are actually folding by our procedure leads to some very interesting new results in number theory. Moreover, the combination of ideas from the paper folding and the number theory produces beautiful, and surprising, new theorems about numbers in *any* base $t \geq 2$, although the folding itself involves only the base 2.

This is a vast topic, but we will restrain ourselves and focus our attention on certain special features of our subject. In Section 1 we show how to fold certain regular N-gons, including some star N-gons,[3] with what we call 2-*period* folding procedures. However, we first discuss our basic construction algorithm called the *FAT-algorithm* and illustrate it by showing how to construct an exact regular convex 8-gon. Next we look at how to construct a regular 7-gon, and it is this problem which involves us in a 2-period folding procedure (as it turns out, we can also use this same tape to fold star $\{\frac{7}{2}\}$- and $\{\frac{7}{3}\}$-gons). The discussion of the general 2-period folding process follows and leads to the 2-period number theory of Section 2. Finally, we observe that a 1-period folding procedure will produce a special class of regular polygons that includes all those whose number of sides is a Fermat number. We illustrate the special case which produces regular 5-gons and we also show how, by introducing secondary crease lines, we obtain tape from which we can construct a regular convex 10-gon. This example explains why our basic problem involves just the construction of regular N-gons for odd $N \geq 3$.

In Section 2 we discuss the number theory which is generated by the obvious mathematical questions suggested by the geometrical constructions introduced in Section 1. However, the number theory inevitably takes on a life of its own, as, of course, it should, so that some of the questions asked, and answered, go well beyond the geometry. We restrain ourselves (again!) and only give a sample of those results. We recommend the reader to consult $[2, 3, 4]$ for further development of the number theory associated with 2-period folding.

---

[3] We will give some hints, and references, that should enable the reader to construct *any* desired star polygon. A *star* $\{\frac{b}{a}\}$-*gon*, sometimes less precisely referred to as a *star b-gon*, is the polygonal path which successively visits every a^{th} vertex of a regular convex b-gon (see [1]).

In Section 3 we abandon the restriction that our basic folding procedure be of period 1 or 2. In this section we provide, first through a concrete example (the construction of a star 11-gon), then in more general terms, a description of the primary folding procedure when N is any given odd integer. As mentioned, and exemplified in Section 1, this is sufficient to enable us to construct a regular N-gon for *any* N, since by the simple addition of the appropriate secondary crease lines we are able to obtain the N-gons with N even.

The instructions for how to fold a regular N-gon, where N is odd, lead us to a remarkable algorithm by which we are able to determine the smallest number ℓ such that $2^\ell \equiv \pm 1 \bmod N$. The algorithm also tells us, for any particular N, which sign to take. This is discussed in Section 4, along with the generalization when 2 is replaced by an arbitrary integer $t \geq 2$, so that N becomes an integer prime to t. The main references for the material of this section are $[4, 5, 6]$.

1. FOLDING 2-PERIOD REGULAR POLYGONS

Before beginning the description of our approximate constructions, let us explain a precise and fundamental folding procedure, involving a straight strip of paper (with parallel edges), that may be used to make the top edge of the strip describe the sides of a regular star $\{\frac{b}{a}\}$-gon (see [1]) where a and b are mutually prime integers with $a < \frac{b}{2}$ (see, for example, Figure 5(c), where $b = 7$ and $a = 2$).[4]

For the moment assume that we have a straight strip of paper (adding machine tape, computer punch tape, or ordinary *unreinforced* gummed tape works well) that has *creases* or *folds* along straight lines emanating from vertices at the top edge of the strip. Further assume that the creases at those vertices, labeled A_{nk}, on the top edge, form identical angles of $\frac{a\pi}{b}$ with the top edge (as shown in Figure 1(a)). Suppose further that these vertices are equally spaced. If we fold this strip on $A_{nk}A_{nk+2}$, as shown in Figure 1(b), and then twist the tape so that it folds on $A_{nk}A_{nk+1}$, as shown in Figure 1(c), the direction of the *top edge* of the tape will be rotated through an angle of $2\left(\frac{a\pi}{b}\right)$. We call this process of *folding and twisting* the **FAT**-algorithm (see $[7, 8]$).

Now observe that if the FAT-algorithm is performed on a sequence of angles, each of which measures $\frac{a\pi}{b}$, at the first b of a number of equally spaced locations along the top of the tape, in our case at A_{nk} for $n = 0, 1, 2, \ldots, b-1$, then the top edge of the tape will have turned through an angle of $2a\pi$, so that the point A_{bk} will then be coincident with A_0. Thus

---

[4] The condition $a < \frac{b}{2}$ is in no way restrictive, since a regular star $\{\frac{b}{a}\}$-gon is completely equivalent to a regular star $\{\frac{b}{b-a}\}$-gon, where the sides of the polygon visit the vertices in the opposite order.

the top edge of the tape will have visited every ath vertex of a bounding regular convex b-gon, and hence determines a regular $\{\frac{b}{a}\}$-gon. (As an example, see Figure 5(c) where $a = 2$ and $b = 7$.)

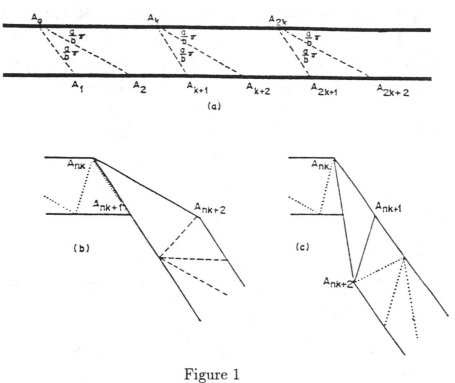

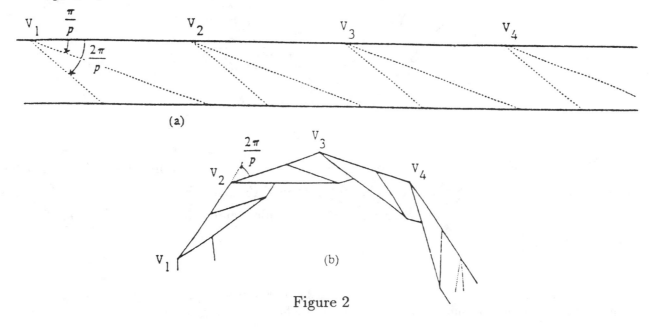

Figure 1

Figure 2 illustrates how a suitably creased strip of paper may be folded by the FAT-algorithm to produce a regular convex p-gon (which may be thought of, for these purposes, as a star $\{\frac{p}{1}\}$-gon).

Figure 2

Let us now illustrate how the FAT-algorithm may be used to fold a regular convex 8-gon. Figure 3(a) shows a straight strip of paper on which the dotted lines indicate certain special exact crease lines. In fact, these crease lines occur at equally spaced intervals along the top of the tape so that the angles occurring at the top of each vertical line are (from left to right) $\frac{\pi}{2}$, $\frac{\pi}{4}$, $\frac{\pi}{8}$, $\frac{\pi}{8}$. Figuring out how to fold a strip of tape to obtain this arrangement of crease lines could be an interesting exercise for students of plane geometry (complete instructions are given in [HP 7]). Our immediate interest is focused on the observation that this tape has, at equally spaced intervals along the top edge, adjacent angles each measuring $\frac{\pi}{8}$, and we can therefore execute the FAT-algorithm at 8 consecutive vertices along the top of the tape to produce the regular convex 8-gon shown in Figure 3(b).[5]

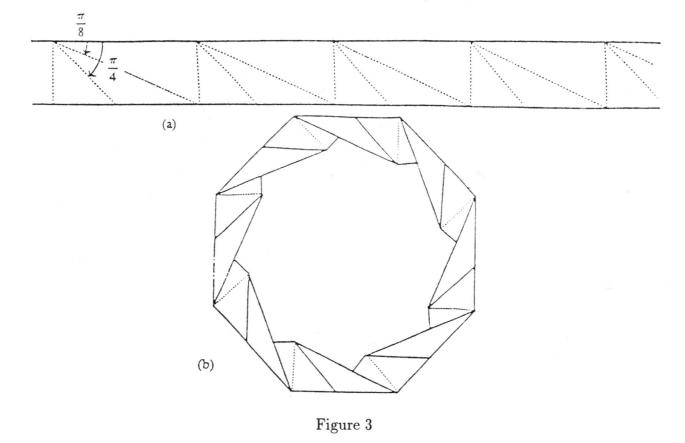

Figure 3

We note, before leaving this example, that the tape shown in Figure 3(a) also has suitable crease lines that make it possible to use the FAT-algorithm to fold a regular convex 4-gon. We leave this as an exercise for the reader and turn to a more challenging construction.

---

[5] Of course, in constructing the model one would cut the tape on the first vertical line and glue a section at the end to the beginning so that the model would form a closed polygon.

Since the regular convex 7-gon is the first polygon we encounter for which we do not have available a Euclidean construction, we are faced with a real difficulty in making available a crease line making an angle of $\frac{\pi}{7}$ with the top edge of the tape. We proceed by adopting a general policy that will prove to be very effective – we call it our *optimistic strategy*. Assume that we *can* crease an angle of $\frac{2\pi}{7}$ (certainly we can come close) as shown in Figure 4(a). Given that we have the angle of $\frac{2\pi}{7}$, it is then a trivial matter to fold the top edge of the strip DOWN to bisect this angle, producing two adjacent angles of $\frac{\pi}{7}$ at the top edge as shown in Figure 4(b) (We say that $\frac{\pi}{7}$ is the *putative* angle on this tape.). Then, since we are content with this arrangement, we go to the bottom of the tape where we observe that the angle to the right of the last crease line is $\frac{6\pi}{7}$ – and we decide we abhor even multiples of π in the numerator of any angle next to the edge of the tape, so we bisect this angle of $\frac{6\pi}{7}$, by bringing the bottom edge of the tape UP to coincide with the last crease line as shown in Figure 4(c).

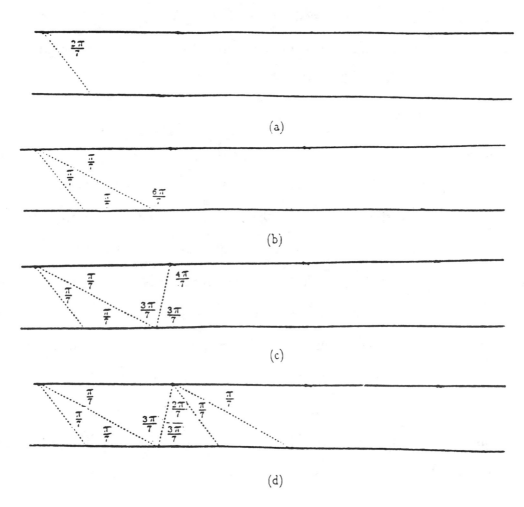

Figure 4

We settle for this (because we decide to be content with an odd multiple of π in the numerator) and go to the top of the tape where we observe that the angle to the right of the last crease line is $\frac{4\pi}{7}$ – and, since we have decided we cannot tolerate an even multiple of π in any angle next to the edge of the tape, we are forced to bisect this angle twice, each time bringing the top edge of the tape DOWN to coincide with the last crease line, obtaining the arrangement of crease lines shown in Figure 4(d).

But now we notice something miraculous has occurred! If we had really started with an angle of exactly $\frac{2\pi}{7}$, and if we now continue introducing crease lines by repeatedly folding the tape DOWN twice at the top and UP once at the bottom, we get precisely what we want; namely, pairs of adjacent angles, measuring $\frac{\pi}{7}$, at equally spaced intervals along the top edge of the tape. Let us call this folding procedure the D^2U^1-*folding procedure* (or, more simply – and especially when we are concerned merely with the related number theory – the $(2, 1)$-folding procedure) and call the strip of creased paper it produces D^2U^1-*tape* (or, again more simply, the $(2, 1)$-tape).

It is clear that the D^2U^1 folding procedure produces tape on which the angles approach the values indicated on Figure 5(a) – otherwise the models would not have turned out as they did. But how do we *prove* that this will be the case? A very direct approach (that works well with very elementary students) would be to admit that the first angle folded down from the top of the tape in Figure 4(a) might not have been precisely $\frac{2\pi}{7}$. Then the bisection forming the next crease would make the two acute angles nearest the top edge in Figure 4(b) only approximately $\frac{\pi}{7}$; let us call them $\frac{\pi}{7}+\epsilon$.[6] Consequently the angle to the right of this crease, at the bottom of the tape, would measure $\frac{6\pi}{7}-\epsilon$. When this angle is bisected, by folding up, the resulting acute angles nearest the bottom of the tape, labeled $\frac{3\pi}{7}$ in Figure 4(c), would measure $\frac{3\pi}{7}-\frac{\epsilon}{2}$, forcing the angle to the right of this crease line at the top of the tape to have measure $\frac{4\pi}{7}+\frac{\epsilon}{2}$. When this last angle is bisected twice by folding the tape down

---

[6] Notice that ϵ may be either positive or negative.

the two acute angles nearest the top edge of the tape will measure $\frac{\pi}{7}+\frac{\epsilon}{2^3}$. This makes it clear that every time we repeat a $D^2 U^1$ folding on the tape the error is reduced by a factor of 2^3.

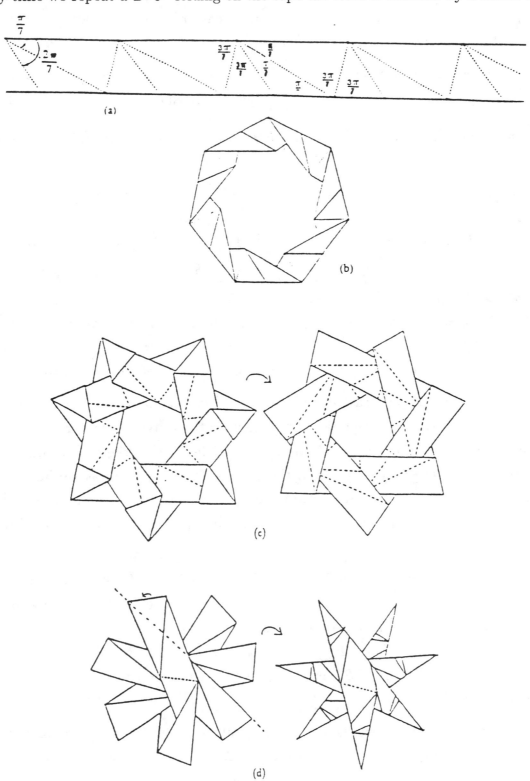

Figure 5

We are now in the position of having many *answerable* questions. Here are just two.

(1) Can we use the same general approach used for folding a convex 7-gon with other values of N, with N odd? Can we prove, in general, that the actual angles on the tape really converge to the putative angle we originally sought?

(2) What happens if we consider general folding procedures with other periods, such as

$$D^3 U^3, \; D^4 U^2, \; \text{or,} \; D^3 U^1 D^1 U^3 D^1 U^1 ?$$

(The *period* is determined by the repeat of the exponents, so these examples have periods 1, 2 and 3 respectively.)

It turns out that the answers to these questions can be obtained by a straightforward use of the following:

LEMMA 1.1 *Let the sequence* $\{x_k\}$, $k = 0, 1, 2, \ldots$ *be generated by the recurrence relation*

$$x_k + ax_{k+1} = b, \quad k = 0, 1, 2, \ldots ; \tag{1.1}$$

then if $|a| > 1$, $x_k \to \frac{b}{1+a}$ *as* $k \to \infty$.

PROOF Set $x_k = \frac{b}{1+a} + y_k$. Then $y_k + ay_{k+1} = 0$. It follows that $y_k = \left(\frac{-1}{a}\right)^k y_0$. If $|a| > 1$, $\left(\frac{-1}{a}\right)^k \to 0$, so that $y_k \to 0$ as $k \to \infty$. Hence $x_k \to \frac{b}{1+a}$ as $k \to \infty$. Notice that y_k is the *error* at the kth stage, so that the absolute value of y_k is equal to $\frac{1}{|a|^k} |y_0|$.

This result is a special case of the well-known contraction mapping principle (see, eg, [9]). We point out that it is significant that neither the convergence nor the limit depends on the initial value x_0. This implies, in terms of the folding, that the process will converge no matter how we fold the tape to produce the first crease line. And, as we have seen in our example, and as we will soon demonstrate in general, the result of the lemma tells us that the convergence of our folding procedure is rapid, since in all cases $|a|$ will be a positive power of 2.

Let us now look at the general 1-period folding procedure $D^n U^n$. A typical portion of the tape would appear as illustrated in Figure 6(a). Then if the folding process had been started with an arbitrary angle u_0 we would have

$$u_k + 2^n u_{k+1} = \pi, \quad k = 0, 1, 2, \ldots \tag{1.2}$$

with $|u_0| < \frac{\pi}{2}$. Equation (1.2) is of the form (1.1), so it follows immediately from Lemma 1.1 that

$$u_k \to \frac{\pi}{2^n + 1} \quad \text{as} \quad k \to \infty. \tag{1.3}$$

Furthermore, we can see that, if the original fold differed from the putative angle of $\frac{\pi}{2^n+1}$ by an error E_0, then the error at the kth stage of the $D^n U^n$-folding procedure would be given by

$$E_k = \frac{E_o}{2^{nk}}.$$

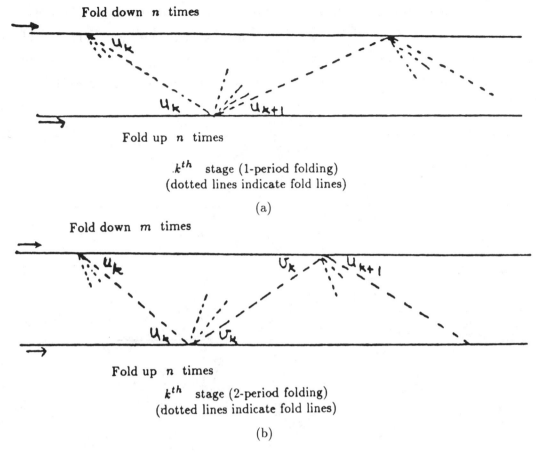

Figure 6

We now see from (1.3) that the $D^n U^n$-folding procedure produces tape from which we may construct regular convex (2^n+1)-gons – and, of course, these include the regular convex N-gons where N is any Fermat number, prime or not. We feel certain that the ancient Greeks and Gauss would have appreciated the fact that when $n = 1$, 2, 4, 8 and 16 the $D^n U^n$-folding procedure produces tape from which we can construct (by means of the FAT-algorithm) the 3-, 5-, 17-, 257-, and 65537-gons, respectively.[7] Moreover, if $n = 3$ we construct the 9-gon, whose non constructibility by Euclidean tools is very closely related to the non trisectibility of the angle.

But there is even more! Because the measure of the smallest acute angle at the bottom and the top of this tape is the same, we may construct the (2^n+1)-gons by special methods – which are, however, not as comprehensive as our FAT-algorithm.

---

[7] Here we could go on at considerable length showing how the folded tape can be used to construct Platonic solids, rotating rings of tetrahedra, triangular and pentagonal dipyramids, collapsoids, . . . Again, we must restrain ourselves and simply refer the interested reader to [8], where very explicit step-by-step instructions are given (along with some of the remarkable mathematics connected with these fascinating polyhedra).

Figure 7 shows how D^2U^2-tape (a) may be folded along just the short lines of the creased tape to form the outline of a small regular convex pentagon (b), and along just the long lines of the creased tape to form the outline of a slightly larger regular convex pentagon (c); and, finally, we show in (d) the regular convex pentagon formed by executing the FAT-algorithm.

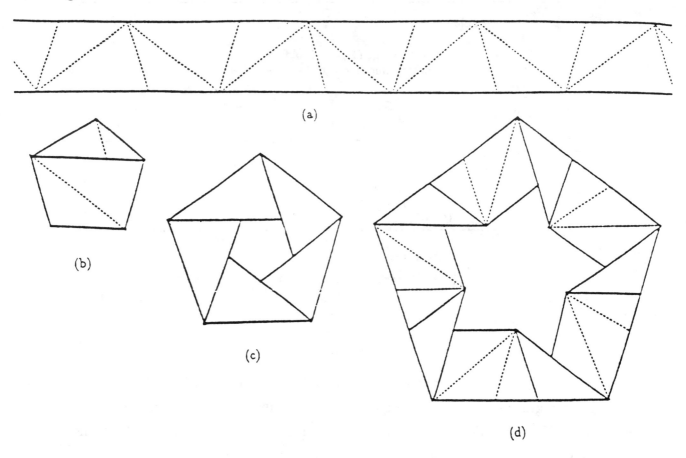

Figure 7

It is now easy to demonstrate how we deal with regular convex polygons with $2^c N$ sides, N odd, if we know how to deal with N-gons. If, for example, we wished to construct a regular 10-gon then we take the D^2U^2-tape and introduce a *secondary crease line* by bisecting each of the angles of $\frac{\pi}{5}$ next to the top edge of the tage. The FAT-algorithm may be used on the resulting tape to produce the regular convex 10-gon, as illustrated in Figure 8. It is now clear how we deal with a 20-gon, a 40-gon, . . .

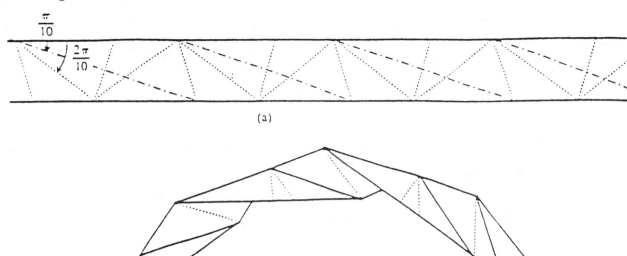

$$\frac{\pi}{10}$$

$$\frac{2\pi}{10}$$

(a)

(b)

Figure 8

Now we turn to the general 2-period folding prodecure, $D^m U^n$. In this case a typical portion of the tape would appear as shown in Figure 6(b). If the folding process had been started with an arbitrary angle $u_0 < \frac{\pi}{2}$ at the top of the tape we would have, at the kth stage,

$$u_k + 2^n v_k = \pi,$$
$$v_k + 2^m u_{k+1} = \pi,$$

and hence it follows that

$$u_k - 2^{m+n} u_{k+1} = \pi(1 - 2^n), \quad k = 0, 1, 2, \ldots \tag{1.4}$$

Thus, again using Lemma 1.1, we see that

$$u_k \to \frac{2^n - 1}{2^{m+n} - 1}\pi \quad \text{as } k \to \infty, \tag{1.5}$$

so that $\frac{2^n - 1}{2^{m+n} - 1}\pi$ is the putative angle. Thus the FAT-algorithm will produce, from this tape, a star $\{\frac{b}{a}\}$-gon,[8] with $b = 2^{m+n} - 1$, $a = 2^n - 1$. By symmetry we infer that

$$v_k \to \frac{2^m - 1}{2^{m+n} - 1}\pi \quad \text{as } k \to \infty. \tag{1.6}$$

Furthermore, if we assume an initial error of E_0 then we know that the error at the kth stage (when the folding $D^m U^n$ has been done exactly k times) will be given by $E_k = \frac{E_0}{2^{(m+n)k}}$. Hence, we see that in the case of our $D^2 U^1$-folding (Figure 4) any initial error E_0 is, as we already saw from our other argument, reduced by a factor of 8 between consecutive stages. It

---

[8] The fraction $\frac{b}{a}$ may turn out not to be reduced; see (2.4).

should be clear why we advise throwing away the first part of the tape – but, likewise, it should also be clear that it is never necessary to throw away very much of the tape. In practice convergence is very rapid indeed, and if one made it a rule of thumb to always throw away the first 20 crease lines on the tape for any iterative folding procedure, it would turn out to be a very conservative rule.

The above technique can be used to prove the convergence for any given folding sequence of general period. This is fine, once we have identified the folding sequence; but the question that is more likely to arise, for the paper-folder, is – *how do we know which sequence of folds to make in order to produce a particular regular star polygon with the FAT-algorithm?* We will return to this question in Section 3, but we first discuss in the next section some of the number theory connected with the 2-period folding procedure.

2. THE NUMBER THEORY OF 2-PERIOD FOLDING NUMBERS

We have observed that the $D^m U^n$-folding procedure, or, more briefly, the (m, n)-folding procedure, produces angles $\frac{\pi}{s}$ on the tape, where

$$s = \frac{2^{m+n}-1}{2^n-1}. \tag{2.1}$$

It is natural, then, to ask when s, given by (2.1), is an integer, and how we can recognize integers of this form. It is not difficult to see intuitively that, in fact, s is an integer if and only if $n \mid m$. However, any attempt to *prove* this quickly reveals that the crucial role played by the base 2 in modeling the folding procedure is not reflected in the argument. In fact, once we move into number theory, we want to get away from this restriction; in doing so *we greatly enrich the theory and enlarge its scope.*

The appropriate setting seems to be the following. We consider a pair of positive integers (t, u) such that $t > u$ and $\gcd(t, u) = 1$. Thus (t, u) generalizes the pair $(2, 1)$ which occurs in our paper-folding procedure, and we say that (t, u) is an *acceptable* pair. Then our first number-theoretical result is the following.

THEOREM 2.1 *Let a, b be positive integers with $\gcd(a, b) = d$. Then, for any acceptable pair (t, u),*
$$\gcd\left(t^a - u^a, t^b - u^b\right) = t^d - u^d.$$

PROOF The conclusion clearly holds if $b \mid a$ or $a \mid b$, so we assume otherwise. Also it is plain that $t^d - u^d$ is a common factor of $t^a - u^a$ and $t^b - u^b$ so it remains to show that, if c is a common factor of $t^a - u^a$ and $t^b - u^b$, then $c \mid (t^d - u^d)$. We express d as $d = ka + \ell b$ where, without real loss of generality, we may assume $k > 0, \ell < 0$. Then
$$u^{-\ell b}(t^d - u^d) = t^d u^{-\ell b} - u^{ka} = t^d(u^{-\ell b} - t^{-\ell b}) + t^{ka} - u^{ka}.$$
Thus if $c \mid (t^a - u^a)$ and $c \mid (t^b - u^b)$ then $c \mid u^{-\ell b}(t^d - u^d)$. But since $\gcd(t, u) = 1$ it is obvious that $\gcd(t^a - u^a, u) = 1$. Hence $\gcd(c, u) = 1$, so that, finally, $c \mid (t^d - u^d)$.

COROLLARY 2.2 $(t^b - u^b) \mid (t^a - u^a) \Leftrightarrow b \mid a$.

PROOF It suffices to observe that $(t^b - u^b) = (t^d - u^d) \Leftrightarrow b = d$. This follows from

PROPOSITION 2.3 $t^x - u^x$ is an increasing function of x in the range $x \geq 0$.

PROOF Let $0 \leq x_1 < x_2$. Then it is easy to see that

$$\frac{t^{x_2} - u^{x_2}}{t^{x_1} - u^{x_1}} > t^{x_2 - x_1} \tag{2.2}$$

and $t^{x_2 - x_1} > 1$ since $t \geq 2$.

COROLLARY 2.4 In general, the rational number $\dfrac{t^a - u^a}{t^b - u^b}$ reduces to the reduced fraction

$$\frac{t^a - u^a}{t^d - u^d} \bigg/ \frac{t^b - u^b}{t^d - u^d},$$

where $d = \gcd(a, b)$.

We prefer at this level not to get involved in discussing *numerals with rational bases*. Thus, to deal with the recognition question we will assume $u = 1$. However, we use the previous reasoning to introduce the set of (t, u)-folding numbers $\mathcal{F}_{tu}$, where

$$\mathcal{F}_{tu} = \left\{ \frac{t^{xy} - u^{xy}}{t^x - u^x} \right\}, \quad x, y \text{ positive integers with } y \geq 2. \tag{2.3}$$

We write $\mathcal{F}_t$ for $\mathcal{F}_{t1}$ and observe that $\mathcal{F}_2$ is precisely the set of numbers s such that we can fold a regular convex s-gon by a 2-period folding procedure (see Figure 9).

Folding numbers; $s = \dfrac{2^{xy} - 1}{2^x - 1} = (x, y)$

An s-gon may be approximated by a $\{ D^{x(y-1)} U^x \}$-folding.

Y / $2^Y - 1$	1	2	3	4	5	6	7	8	9	10	11	x
20	1048575											
19	524287											
18	262143											
17	131071											
16	65535											
15	32767											
14	16383											
13	8191	22369621										
12	4095	5592405										
11	2047	1398101										
10	1023	349525										
9	511	87381	19173961									
8	255	21845	2396745									
7	127	5461	299593	17895697								
6	63	1365	37449	1118481	34636833							
5	31	341	4681	69905	1082401	17043521						
4	15	85	585	4369	33825	266305	2113665	16843009				
3	7	21	73	273	1057	4161	16513	65793	262657	1049601	4196353	16781313
2	3	5	9	17	33	65	129	257	513	1025	2049	$2^X + 1$
1	1	1	1	1	1	1	1	1	1	1	1	1

Figure 9

Now suppose N given. Let us write N in base t. Then we see from (2.3) that $N \in \mathcal{F}_t$ if and only if, in base t,

$$N = 10\ldots0\,10\ldots0\,\ldots\,10\ldots01 = \frac{t^{xy}-1}{t^x-1},$$

where the repeating piece $10\ldots0$ consists of 1 followed by $(x-1)$ zeros, and there are y 1's. Thus we not only recognize the members of $\mathcal{F}_t$ but can also assign to each $N \in \mathcal{F}_t$ its (x, y)-coordinates.

EXAMPLE 2.1 Suppose we wish to fold a regular 73-gon. We write, in base 2,

$$73 = 1001001.$$

Thus $x = 3$, $y = 3$, so $73 = \frac{2^9-1}{2^3-1}$. Comparing this with (2.1), we see that we may fold a regular convex 73-gon by the D^6U^3- or $(6, 3)$-folding procedure.

What star polygons may we fold by a 2-period folding procedure? The answer is provided by Corollary 2.4. For we see that

$$\frac{t^a-1}{t^b-1} = \frac{t^a-1}{t^d-1} \Big/ \frac{t^b-1}{t^d-1}, \quad \text{the ratio of two numbers[9] in } \mathcal{F}_t \tag{2.4}$$

and, in base t, $\frac{t^a-1}{t^d-1}$ and $\frac{t^b-1}{t^d-1}$ have the same repeating piece but mutually prime y-lengths. Such then are the rational numbers expressible as $\frac{t^a-1}{t^b-1}$ for some a, b. Suppose we confine attention to the case $a > b$, and ask what star $\left\{\frac{N}{N'}\right\}$-gons, with N' odd, we may fold by a 2-period folding procedure, simply using the smallest angle at the top of the tape.[10] Our argument shows that we will require $N, N' \in \mathcal{F}_2$, where N' has the same repeating piece as N and a y-length prime to that of N. In particular, we see that we may only fold a $\left\{\frac{N}{N'}\right\}$-gon by a 2-period procedure if we may fold a convex N-gon, that is, if N is a folding number.

EXAMPLE 2.2 We have seen that the $(6, 3)$-folding procedure folds a convex 73-gon. What star 73-gons can we fold and by what folding procedures? Now (see (2.4)) we have

$$\frac{2^a-1}{2^b-1} = \frac{2^a-1}{2^d-1} \Big/ \frac{2^b-1}{2^d-1} \quad \text{and} \quad a = 9, \ d = 3, \quad \text{since} \ \frac{2^a-1}{2^d-1} = 73.$$

The only available value for b is $b = 6$ (recall that we require $b < a$, $b > d$, $\gcd(a, b) = d$). Thus we can fold a $\left\{\frac{73}{9}\right\}$-gon using a $(3, 6)$ folding procedure; but this implies that we may instead use the $(6, 3)$ procedure and use the *bottom* of the tape instead of the top. That is, since the angle of $\frac{9\pi}{73}$ will appear at the bottom of the D^6U^3-tape, it would be required to flip the tape so that this angle appeared at the top — and then simply apply

---

[9] More precisely, we should speak of the ratio of two numbers in $\overline{\mathcal{F}}_t = \mathcal{F}_t \cup \{1\}$. It is often convenient to regard 1 as a degenerate folding number; more generally, we form $\overline{\mathcal{F}}_{tu}$ by adjoining 1 to $\mathcal{F}_{tu}$.

[10] In fact, each different kind of crease line on the tape will produce *some* star polygon. Of course, if the crease line used to fold the star polygon is not one of those next to the top or bottom of the tape then N' is an *even* number.

the FAT-algorithm on the crease lines nearest the top edge of the tape. This, then, is the only star $\{\frac{N}{N'}\}$-gon we can fold with a 2-period folding procedure with $N = 73$ and N' odd.[11]

EXAMPLE 2.3 In our previous example we were actually able to fold the $\{\frac{73}{9}\}$-gon by using the same tape as that used to fold the convex 73-gon (merely turning it upside down). It will not always be the case that, when an $\{\frac{N}{N'}\}$-gon can be folded by a 2-period folding procedure, we may use the *same* tape as that used to fold the convex N-gon. Thus, if we ask what star 1023-gons we may fold by a 2-period procedure, we express 1023 as $\frac{2^{10}-1}{2^1-1}$, so the answer is that we may fold a $\left\{\frac{2^{10}-1}{2^b-1}\right\}$-gon, where $1 < b < 10$ and $\gcd(10, b) = 1$. Thus $b = 3, 7$, or 9. The associated folding procedures are D^7U^3, D^3U^7, D^1U^9, producing the $\{\frac{1023}{7}\}$-gon, the $\{\frac{1023}{127}\}$-gon, and the $\{\frac{1023}{511}\}$-gon, respectively. Of course, the last could be obtained by flipping the D^9U^1-tape which produced the convex $\{1023\}$-gon, just as the $\{\frac{1023}{127}\}$-gon could be obtained by flipping the D^7U^3-tape.

There is much more to be said about the number theory inspired by 2-period paper folding (see [3, 4, 10]). We will only mention here the following striking new result which should be compared with Theorem 2.1.

THEOREM 2.5 *Let a, b be positive integers with $\mathrm{lcm}(a, b) = \ell$. Then, for any acceptable pair (t, u),*

$$\mathrm{lcm}\left(t^a - u^a, \ t^b - u^b\right) = t^\ell - u^\ell \ \Leftrightarrow \ a/b \ or \ b/a.$$

For a proof see [10].

3. GENERAL FOLDING PROCEDURES

We now answer the question we posed at the end of Section 1. Let us demonstrate our approach with an example, which is sufficiently general to show the way to the construction of any regular star $\{\frac{b}{a}\}$-gon, where a, b are mutually prime odd integers with $a < \frac{b}{2}$.

Suppose we want to construct a regular star $\{\frac{11}{3}\}$-gon. Then, of course, $b = 11$, $a = 3$ and we proceed as we did when we wished to construct the regular convex 7-gon (in Section 1) – we adopt our *optimistic strategy*. We assume we can fold the desired putative angle of $\frac{3\pi}{11}$ at A_0 and we adhere to the same principles that we used in constructing the regular 7-gon, namely,

(1) Each new crease line goes in the forward (left to right) direction along the strip of paper.
(2) Each new crease line always *bisects* the angle between the last crease line and the edge of the tape from which it emanates.

---

[11] The other star polygons that may be folded from this same tape, dropping the requirement that N' is odd, are the

$$\{\tfrac{73}{2}\}\text{-}, \ \{\tfrac{73}{4}\}\text{-}, \ \{\tfrac{73}{8}\}\text{-}, \ \{\tfrac{73}{16}\}\text{-}, \ \{\tfrac{73}{32}\}\text{-}, \ \{\tfrac{73}{18}\}\text{-}, \ \text{and} \ \{\tfrac{73}{36}\}\text{- gons.}$$

(3) The bisection of angles at any vertex continues until a crease line produces an angle of the form $\frac{a'\pi}{b}$ where a' is an *odd* number; then the folding stops at that vertex and commences at the intersection point of the last crease line with the other edge of the tape.

Following this procedure will, in fact, result in tape whose angles converge to those shown in Figure 10(b). We could denote this folding process as $D^1 U^3 D^1 U^1 D^3 U^1$, interpreted in the obvious way on the tape — that is the first exponent "1" refers to the one bisection (producing a line in a downward direction) at the vertices A_{6n} (for $n = 0, 1, 2, \ldots$) on the top of the tape; similarly, the "3" refers to the 3 bisections (producing creases in an upward direction) made at the bottom of the tape through the vertices A_{6n+1}; etc. However, since the folding process is *duplicated* halfway through, we can abbreviate the notation and write simply $\{1, 3, 1\}$, with the understanding that we alternately fold from the top and bottom of the tape as described, with the *number* of bisections at each vertex running, in order, through the values $1, 3, 1, \ldots$ We call this a *primary folding procedure of period* 3 or a *3-period folding.*

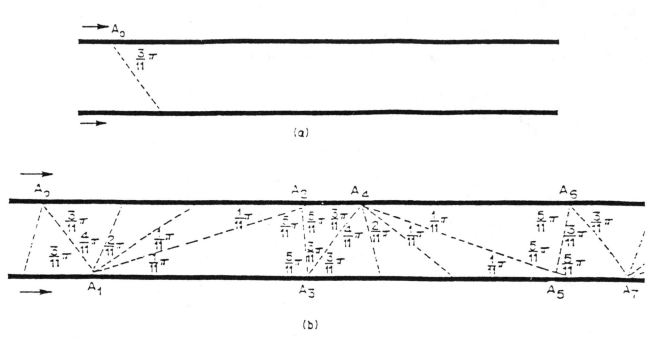

(Note that the indexing of the vertices is <u>not</u> the same as that in Figure 1.)

Figure 10

The proof of convergence for the general folding procedure (see below) is similar to that for the primary folding procedure of period 2. We leave the details to the reader, and explore here what we can do with this $\{1, 3, 1\}$-tape. First, note that, starting with the putative angle $\frac{3\pi}{11}$ at the top of the tape, we produce a putative angle of $\frac{\pi}{11}$ at the bottom of the tape, then a putative angle of $\frac{5\pi}{11}$ at the top of the tape, then a putative angle of $\frac{3\pi}{11}$ at the *bottom* of the tape, and so on. Hence we see that we could use this tape to fold a star

$\{\frac{11}{3}\}$-gon, a convex 11-gon, and a star $\{\frac{11}{5}\}$-gon. More still is true; for, as we see, if there are crease lines enabling us to fold a star $\{\frac{11}{a}\}$-gon, there will be crease lines enabling us to fold star $\left\{\frac{11}{2^k a}\right\}$-gons, where $k \geq 0$ takes all values such that $2^{k+1}a < 11$. These features, as described for $b = 11$, apply for any odd number b. However, this tape has a special symmetry as a consequence of its *odd* period; namely that if it is "flipped," about the horizontal line half way between its parallel edges, the result is a *translate* of the original tape. As a practical matter this special symmetry of the tape means that we can use either the top edge or the bottom edge of the tape to construct our polygons. On tapes with an *even* period the top edge and the bottom edge of the tape are not translates of each other (under the horizontal flip), which simply means that care must be taken in choosing the edge of the tape used to construct a specific polygon. Figures 11(a) and 11(b) show the completed $\{\frac{11}{3}\}$-, $\{\frac{11}{4}\}$-gons, respectively.

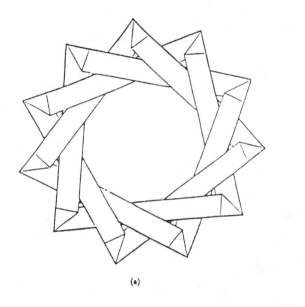
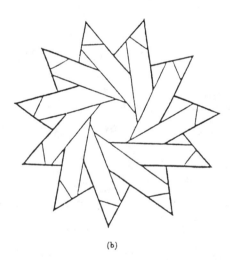

(a)   (b)

Figure 11

Now, to set the scene for the number theory of Section 4, let us look at the patterns in the *arithmetic* of the computations when $a = 3$ and $b = 11$. Referring to Figure 10(b) we observe that

the smallest angle to the right of A_n where $n =$	is of the form $(a/11)\pi$ where $a =$	and the number of bisections at the next vertex $=$
0	3	3
1	1	1
2	5	1
3	3	3
4	1	1
5	5	1

We could write this in shorthand form as follows:

$$(b =) \ 11 \quad \begin{vmatrix} (a=) \ 3 & 1 & 5 \\ 3 & 1 & 1 \end{vmatrix} \qquad (3.1)$$

Observe that, had we started with the putative angle of $\frac{\pi}{11}$, then the *symbol* (3.1) would have taken the following form.

$$(b =) \ 11 \quad \begin{vmatrix} (a=) \ 1 & 5 & 3 \\ 1 & 1 & 3 \end{vmatrix} \qquad (3.1')$$

In fact, it should be clear that we can *start anywhere* (with $a = 1$, 3 or 5) and the resulting symbol, analogous to (3.1), will be obtained by cyclic permutation of the matrix component of the symbol, placing our choice of a in the first position along the top row.

In general suppose we wish to fold a $\{\frac{b}{a}\}$-gon. Then we may construct a *2-symbol* as follows. Let us write

$$b \quad \begin{vmatrix} a_1 & a_2 & . & . & . & a_r \\ k_1 & k_2 & . & . & . & k_r \end{vmatrix} \qquad (3.2)$$

where b, a_i ($a_1 = a$) are odd, $a_i < \frac{b}{2}$, and

$$b = a_i + 2^{k_i} a_{i+1}, \quad i = 1, 2, \ldots, r, \quad a_{r+1} = a_1. \qquad (3.3)$$

We comment that, given any two odd numbers a and b, with $a < \frac{b}{2}$, there is always a completely determined unique symbol (3.2) with $a_1 = a$; this is proved in [4, 6]. At this stage, we do not assume that $\gcd(b, a) = 1$, but we do assume that the list $a_1, a_2, \ldots, a_r$ is without repeats. Indeed, if $\gcd(b, a) = 1$ we say that the symbol (3.2) is *reduced*, and if there are no repeats among the a_i's we say that the symbol (3.2) is *contracted*.[12] We regard (3.2) as encoding the general folding procedure to which we have referred.

4. THE NUMBER THEORY OF THE GENERAL FOLDING PROCEDURE

In this section we give a very much abbreviated account of the number theory associated with, or suggested by, the general folding procedure. We give no proofs here; they are to be found in [4, 6, 10].

We recall the symbol

$$b \quad \begin{vmatrix} a_1 & a_2 & . & . & . & a_r \\ k_1 & k_2 & . & . & . & k_r \end{vmatrix} \qquad (4.1)$$

---

[12] It is, of course, theoretically possible to consider 2-symbols (3.2) in which repetitions among the a_i are allowed.

introduced in the previous section; the passage from a_i to a_{i+1} ($a_{r+1} = a_1$) is achieved by means of the rule

$$\psi(a) = a', \text{ where } b = a + 2^k a', \text{ with } k \text{ maximal}. \tag{4.2}$$

Our first theorem guarantees that, as claimed above, there always *is* a symbol (4.1).

THEOREM 4.1 *Let* $b \geq 3$ *be odd, and let* S *be the set of odd positive integers* $a < \frac{b}{2}$. *Then* ψ, *given by* (4.2), *is a permutation of* S *which preserves* $\gcd(b, a)$.

We next announce an important theorem, which shows that our folding procedure has led to a very satisfying algorithm in number theory.

THEOREM 4.2 *Let the symbol* (4.1) *be reduced and contracted. Then, if* $k = \sum k_i$,
$$2^k \equiv (-1)^r \bmod b.$$
Indeed, k *is the quasi-order of* $2 \bmod b$, *that is, the smallest positive integer* m *such that* $2^m \equiv \pm 1 \bmod b$.

Knowing that $b \mid \left(2^k - (-1)^r\right)$, it is natural to ask for the complementary factor, that is, to try to find b' such that $2^k - (-1)^r = bb'$. We now give an algorithm to achieve this.

We revert to the symbol (4.1) and set $k = \sum k_i$. If we solve the equations
$$a_i + 2^{k_i} a_{i+1} = b, \quad i = 1, 2, \ldots, r \quad (a_{r+1} = a_1) \tag{4.3}$$
in the 'unknowns' $a_1, a_2, \ldots, a_r$, we obtain the solutions (see [11])
$$Ba_i = bA_i, \tag{4.4}$$
where
$$\left. \begin{array}{l} B = 2^k - (-1)^r \\ A_i = 2^{k-k_i-1} - 2^{k-k_i-1-k_{i-2}} + \ldots + (-1)^r 2^{k_i} - (-1)^r, \quad i = 1, 2, \ldots, r \end{array} \right\} \tag{4.5}$$
(Here we find it convenient to think of the suffixes i running over the residue classes $\bmod r$ so that, for example, $k_r = k_0$.) We conclude that
$$2^k - (-1)^r = b \frac{A_i}{a_i}, \text{ any } i, \ 1 \leq i \leq r, \tag{4.6}$$
giving us the complementary factor b', namely, $\frac{A_i}{a_i}$, for any i.

EXAMPLE 4.1 One verifies that

$$641 \ \left| \ \begin{array}{ccccccccc} 1 & 5 & 159 & 241 & 25 & 77 & 141 & 125 & 129 \\ 7 & 2 & 1 & 4 & 3 & 2 & 2 & 2 & 9 \end{array} \right|$$

is a (reduced) symbol. This tells us that the quasi-order of $2 \bmod 641$ is 32 and, indeed, that $2^{32} \equiv -1 \bmod 641$. Thus we have proved Euler's result that the Fermat number $F_5 = 2^{2^5} + 1$ is not prime, for we have one factor, 641. To find the complementary factor we apply (4.6) with $i = 1$. Thus

$$2^{2^5} + 1 = 641 A_1,$$

where, by (4.5),

$$A_1 = 2^{23} - 2^{21} + 2^{19} - 2^{17} + 2^{14} - 2^{10} + 2^9 - 2^7 + 1 = 6700417,$$
$$2^{32} + 1 = 641 \times 6700417.$$

Our arguments have a beautiful interpretation in paper-folding. Suppose we adopt the folding procedure $(k_1, k_2, \ldots, k_r)$; what polygons can we thereby fold? The answer is obtained as follows:

(i)   calculate B and A_i from (4.5);

(ii)   find $q = \gcd(B, A_i)$ and set $B = qb$, $A_i = qa_i$;

(iii)   we may then use the FAT-algorithm, on the top or bottom of the tape, to fold $\{\frac{b}{a_i}\}$-gons, $i = 1, 2, \ldots, r$.

Of course, we may also fold $\{\frac{b}{2^r a_i}\}$-gons, provided $2^r a_i < \frac{b}{2}$.

EXAMPLE 4.2 Suppose we adopt the folding prodecure $(1, 1, 1, 2, 4)$. Then $B = 2^9 + 1 = 513$, and $A_1 = 2^5 - 2^3 + 2^2 - 2 + 1 = 27$. Now $\gcd(513, 27) = 27$, so $b = 19$, $a_1 = 1$. We start with these b, a_1 and construct the symbol

$$19 \quad \begin{vmatrix} 1 & 9 & 5 & 7 & 3 \\ 1 & 1 & 1 & 2 & 4 \end{vmatrix}$$

so that we may fold the convex 19-gon and the star polygons $\{\frac{19}{3}\}$, $\{\frac{19}{5}\}$, $\{\frac{19}{7}\}$, $\{\frac{19}{9}\}$. Notice that, in the interests of arithmetical simplicity, we slightly modified the algorithm we described before the example – having calculated b and a_1, it is certainly quicker to use these and (4.8) to obtain the complete reduced symbol, rather than to calculate the remaining A_i from (4.10).

The Quasi-order Theorem 4.2 may be generalized by replacing 2 by an arbitrary positive integer $t \geq 2$. We describe this fairly sophisticated generalization in [4, 6, 10], and commend it to the diligent reader.

BIBLIOGRAPHY

[1]   Coxeter, H. S. M., *Regular Polytopes*, Methuen (1928).

[2]   Hilton, Peter, and Jean Pedersen, "Folding regular star polygons and number theory," *Math. Intelligencer* 7, No. 1 (1985), 15 - 26.

[3]   Hilton, Peter, and Jean Pedersen, "On a generalization of folding numbers," *Southeast Asian Bulletin of Mathematics*, 12, No. 1 (1988), 53 - 63.

[4]   Hilton, Peter, and Jean Pedersen, "Geometry in Practice and Numbers in Theory," *Monographs in Undergraduate Mathematics*, Vol. 16 (1987), 37pp. (Available from Department of Mathematics, Guilford College, Greensboro, North Carolina 27410).

[5]   Hilton, Peter, and Jean Pedersen, "Certain algorithms in the practice of geometry and the theory of numbers," *Publ. Sec. Mat. Univ. Autonoma Barcelona* **29**, No. 1 (1985), 31 - 64.

[6]   Hilton, Peter, and Jean Pedersen, "The general quasi-order algorithm in number theory," *Int. Journ. Math. and Math. Sci.*, **9**, No. 2 (1986), 245 - 252.

[7]   Hilton, Peter, and Jean Pedersen, "Approximating any regular polygon by folding paper; An interplay of geometry, analysis and number theory," *Math. Mag.*, **56** (1983), 141 - 155.

[8]   Hilton, Peter, and Jean Pedersen, *Build Your Own Polyhedra*, Addison-Wesley, Menlo Park, CA (1987), 175 pp.

[9]   Carroll O. Wilde, *The Contraction Mapping Principle*, UMAP Unit 326 (Lexington, MA: COMAP, Inc.).

[10] Hilton, Peter, and Jean Pedersen, "Folding regular polygons and number theory," *Mathematics in Education*, edited by Themistocles M. Rassias, University of LaVerne, CA (1992), 17-50.

[11] Hilton, Peter, and Jean Pedersen, "On the complementary factor in a new congruence algorithm," *Int. Journ. Math. and Math. Sci.*, **10**, No. 1 (1987), 113 - 123.

The Meanings of Mathematical Proof:
On Relations Between History and Mathematical Education

Evelyne Barbin
Université du Maine
72017 Le Mans, FRANCE

Why is the history of mathematics of interest in mathematical education? First, a pedagogical aspect; the history of mathematics helps a student grasp the meaning and the sense of the mathematical concepts and theories. Studying the history of mathematics allows one to study the construction of mathematical knowledge and to study mathematical activity – to analyze the role of problems, of proof, of conjecture, of evidence, of error. Second, a cultural aspect: studying the history of mathematics enables one to study the scientific, philosophical, cultural and social context in which mathematical knowledge was elaborated.

So, the history of mathematics is a way to change the image of mathematics in the teacher's and in the pupil's head; to see mathematics not as a language or as a spectacle, but as an activity, a human activity. This role is very important in the context of mathematical education in France, because studying the history of mathematics is a way to struggle against a formalist and dogmatic vision of mathematics.

I'm going to illustrate all these aspects with mathematical proof. Why prove? What is the meaning of mathematical proof? These two questions are at the center of any epistemological reflection on the teaching of "proving."

In France, teachers see mathematical proof as a synonym for deductive reasoning. They conceive of the idea of "proof" only as a means to convince. The history of mathematics shows that the idea of mathematical proof has changed as time passed, and that to identify mathematical proof with deductive reasoning hides the questions, the doubts, and the process of invention. I'm going to consider the beginning of "proof" and two important breaks in the seventeenth and nineteenth centuries. I'm going to begin by studying an historical and pedagogical example.

An Historical and Pedagogical Example

A teacher told me about an experience with her pupils. They were fifteen years old and they were studying a famous result in book ten of Euclid's *Elements* in which Euclid establishes that the diagonal and the side of a square are incommensurable. Euclid assumes that in square $ABCD$, AB and BD are commensurable – that there exist two relatively prime numbers m and n such that $AB/BD = m/n$. From the Pythagorean Theorem $m^2 = 2n^2$. The proof follows consideration of the parity of m and n. If m^2 is even then so is m, so n must be odd (since m and n had no common factors). But if m is even then n^2 is even and so is n. The contradiction is that n must be both odd and even.

After obtaining this contradiction the teacher asked the pupils, "What can one conclude from this?" Their conclusion was "The Pythagorean Theorem is false!" Now, we must try to understand the pupils' answer. They certainly understood the idea of proof by contradiction, ergo something must be false. There are three possibilities; first, "the diagonal and side are commensurable" might be false – second, "the Pythagorean Theorem" might be false – third, the argument about odds and evens might be incorrect.

For the pupils it seems evident that two segments have a common measure; it is only necessary to find one that is sufficiently small. For a mathematician there is only one statement that is in doubt, it is the assumption that the diagonal and side are commensurable. The Pythagorean Theorem is the result of a proof which has to be taken as a certainty. We expect the pupils to admit this validity of the Pythagorean Theorem. But, in fact, this assumes a certain rationality of belief of the pupils. For them the commensurability of the two segments is more certain than the Pythagorean Theorem.

Two years ago, after reading the proof by Euclid with first year Arts students, I asked them, "What do you think about this?" They were dissatisfied with the proof. The argument upon which it depends, a number cannot at the same time be both odd and even, was too distant from the original problem of attempting to find a common measure for the side and the diagonal of a square. Contrary to the expectations of the students, the proof avoids all mention of the comparison of line segments. The proof by Euclid seemed artificial.

So, proofs may convince us yet produce no understanding. We can state the result, yet, as Bachelard wrote: "...asserting a result is not enough to grasp its meaning."[1] The proof by using an odd/even argument is typical, in some respects, of proof in Greek geometry. I'm now going to consider the historical evolution of proof and the meaning of proof in Greek geometry.

The Idea of Proof in Greek Geometry

The sixth century B.C. saw the birth of rational and geometrical thought in Ancient Greece. What does "rational and geometrical" thought mean? Here is an example from astronomy. For Anaximandre, if the Earth did not fall, it was not thanks to Zeus, but because it was at the same distance from every point on the celestial circumference. It had no reason to go to the left or to the right, to go up or go down. Why the birth of this rational and geometrical thought? It was not a miracle, as the French philosopher Jean-Pierre Vernant shows.[2] Remember that the sixth century B.C. was also the birth of the city state and democracy. All the business of the city was the object

---

[1] Bachelard, *Le rationalisme appliqué*, p.11

[2] Vernant, *Mythe et pensée chez les Grecs*.

of public disputation using reasoned arguments, what the Greeks called "logos." In the city it was necessary to convince using speech based on reason.

This was also true for metaphysical and scientific subjects. In this way, Plato wrote dialogues in which Socrates had to convince his opponent. For Aristotle, to know scientifically was to know by means of a proof.[1] The Eleatic philosophy was based on the art of public speaking in which the indirect process was appreciated because it led the other speaker to a contradiction.[2]

How is this method used to convince someone? First you say, "Do you accept that 'this' and 'this' is true?" Some points must be admitted. They are called postulates and common notions in Euclid's *Elements*. Then the axiomatic-deductive system response is, "Now, if you accept 'this' and 'this' you must accept 'this'..." For Aristotle, the acceptance of axioms did not rest upon mathematics, but upon metaphysics. They compelled recognition thanks to collective human reasoning.

The odd/even argument bears all the hallmarks of Greek deductive reasoning. The objections to this argument are that it avoids any mention of comparison, of continuum, or of infinity. But, as we have noted, proof is a social act among a group of listeners sharing the same rationality.

Euclid preferred indirect proof to anthyphairesis (see Fowler, "Ratio in early Greek Mathematics"). This process, however, can be directly tied to proposition 11 of Book X of the *Elements* which establishes the nature of incommensurables. Successive subtractions of two magnitudes are said to be incommensurable if there always remains a remainder. The proof is as follows. Consider a square $ABCD$ with diagonal BD. Let E lie on BD so that $DE = DC$, and let F be on BC so that EF is perpendicular to BD. It is easy to show that $BE = EF = FC$. So, we have the successive subtractions: $BD - DC = BE$, and $DC - BE = BC - FC = BF$. To continue, we construct the square $BEFG$. Let H be a point on BF so that $HF = BE$, and let I be on BE so that IH is perpendicular to BF. Then we have the successive subtractions $BF - BE = BH$, and $BE - BH = BI$. Now we construct the square $IHBJ$ and repeat the subtractions [Figure 1]. Since this process can be continued indefinitely, and there is always a remainder, the diagonal and side of a square are incommensurable directly in accordance with proposition 11 of Book X of the *Elements*.

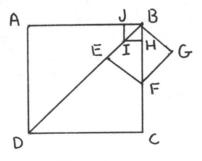

Figure 1

---

[1] Aristotle, *Organon. Seconds Analytiques*, 12, 15.

[2] Szabo, *Greek dialectic and Euclid's axiomatics*.

The problem is that this proof relies upon an infinite process! It is, therefore, open to objection on that account. It may, however, be assumed that the mathematicians of the seventeenth century would have preferred this proof by anthyphairesis to the even/odd argument. In fact, the proof by anthyphairesis offers a construction that is an exact parallel with the idea of incommensurability and can help us understand its nature. The seventeenth century view would be that the one enlightens while the other only convinces.

Proving: A Means to Convince or to Enlighten

The seventeenth century marked a major break in the concept of the understanding of proof. There was widespread dissatisfaction with the proofs of the Ancients. In 1674, Arnauld and Nicole, both jansenists and friends of Pascal, listed their objections to Euclid in *La logique ou l'art de penser* [The logic or the art of thinking].[1]

1. More concerned with certitude than evidence, more concerned with convincing the mind than with enlightening it.

All their other objections derive from this.

2. Proving things that do not need proof. For instance, the sum of two sides of a triangle is greater than the third.
3. Proof by impossibility. Such proofs convince the mind without enlightening it "...because our mind is dissatisfied if it knows something is, but not why it is."
4. Proofs by ways that are distant from the original problem. "All Euclid is full of such proofs by extraneous ways."

The fifth and sixth objections are concerned with methods of proof.

5. Not mindful of the true order of nature.
6. Does not use divisions and partitions.

In Rule IV of his *Règles pour la direction de l'esprit* [Rules for the Direction of the Mind], Descartes explained his dissatisfaction with the texts of the Ancients. The results were obtained by rigorous reasoning, but the Ancients didn't explain *why* these results were true, and how they obtained them. The Ancients gave sophisticated proofs, but they lacked the art of invention - the "true mathematic" – which allows resolution of new problems. Descartes wrote that the Ancients surely were in possession of this art, but they hid it by avarice.[2]

To further understand these criticisms, I'm going to look at the method of proof of Archimedes about the quadrature of surfaces.

---

[1] Arnauld and Nicole, *La Logique ou l'art de penser*, pp. 325 to 331.

[2] Descartes, *Règles pour la direction de l'esprit*, pp. 21 to 25.

Quadratures: The Method of Archimedes

Archimedes proves the equality of two surfaces by establishing the contradiction of both of the inequalities. So, to prove the equality of the area of a circle s_c and the area of a rectangular triangle s_t whose sides are the radius and the circumference of the circle, Archimedes showed that the two inequalities $s_c > s_t$ and $s_c < s_t$ lead to contradictions.

For instance, when the area of the circle is supposed to be superior to the area of the triangle, Archimedes showed that there exist regular polygons inscribed in the circle whose areas are in one instance inferior and in the other instance superior to the area of the triangle.[1] This proof has the advantage of avoiding any consideration about infinity and the composition of a continuum. So we are convinced by the result and we cannot object to it. But the geometers of the seventeenth century criticized it on three grounds:

- it was an indirect method that presupposed knowledge of the result
- it was difficult and required long processes
- each quadrature had to be evaluated case by case.

The archimedean proof lacks an explanation of its invention. The geometers of the seventeenth century thought that Archimedes hid his method. Now we know that Archimedes possessed a method of invention for his proofs, but unfortunately the treatise in which he explained it was only discovered at the beginning of our century in a library in Constantinople.

Legitimate Proof and Methods of Invention

The geometers of the seventeenth century elaborated many methods of invention for their results. These included methods for finding tangents and for finding the center of gravity of an object. But can these results be considered as proven? These methods were widely used in the seventeenth century because they were seen by all as marvelous tools, but their value as proof was not unanimously accepted.

An explanation of the method of invention of a result enlightens because it shows the manner by which the result has been obtained. The Italian Nardi wrote, "All evidence is certain, but all certainty is not evident." This is a significant remark, since from the moment that the geometry of the seventeenth century stepped outside the bounds of deductive reasoning, it was certainty that had to be appealed to as a claim for proof.

In his *Discours de la méthode* (1636), Descartes proposed a methodical enumeration of knowledge which is able to produce evident and certain results. Such an enumeration is used in *La géométrie* to state a general method for resolving a geometrical problem by means of algebraic equations. And when Descartes came to the matter of confidence in his results he wrote, "All things

---

[1] Archimedes, Traité de la mesure du cercle, *Oeuvres complètes,* pp. 127 to 128.

we conceive very clearly and very distinctly to be true."[1] If a method leads to evidence and to certainty, then it can lay claim to being of value as proof. We are obliged to accept it as a second method of proof alongside that of deductive reasoning. Descartes dealt with this in the *Réponses aux secondes objections* [Answers to the Second Objections] of his *Méditations métaphysiques* [Metaphysical Meditations].

> "The method of proof is twofold: one by analysis or resolution, the other by synthesis or composition.   Analysis shows the true way by which something has been methodically invented ... such that, if the reader wishes to follow it ... he will not be able to understand it as perfectly, and will render it less his own, than if he himself had invented it.   Synthesis ... derives from a long list of definitions, demands, axioms, theorems, and problems ... These drag along the consent of the reader, however obstinate or opinionated he may be; but they do not give, as does the other, complete satisfaction to the minds of those who wish to learn, because they do not teach the method by which the thing has been invented."[2]

The idea of two means of proof, analysis and synthesis, was often to recur in the succeeding centuries, particularly with respect to the learning of mathematics. Clairaut wrote in his *Eléments de géométrie* (1765) that beginners should be shown the way by which things were being discovered.[3] Up to the end of the nineteenth century, authors rejected the idea of proof as deductive reasoning when it was a question of enlightening beginners. Might it not be that a feeling for proof derives from the very act of discovering a result? This remark has important pedagogical implications to which I'm going to return.

The geometers of the seventeenth century leaned heavily on evidence. It was, for them, a source of certainty. Arnauld and Nicole were critical of those who set out to prove what was clear and evident. Legendre in his *Eléments de géométrie* (1823) defined a theorem as "... a truth which becomes evident by means of a reasoning called proof." The ideas of proof and evidence are strongly linked. However, a contemporary of Legendre, Bolzano, wrote a memoir in which he strongly opposed the idea of proof as the production of evidence. This marked a break in the meaning of proof which was amplified throughout the development of non-Euclidean geometries and the formalist movement.

Mathematical proof, evidence and contradiction

In the preface to his memoir of 1817 whose long title starts with *Rein Analytisher Beweis* [Pure Analytical Proof] Bolzano explained why he considered the existing proofs of what is now known as the fundamental theorem of algebra as unsatisfactory. In its modern form this theorem states that every polynomial over the complex numbers can be factored with linear factors. Some

---

[1] Descartes, *Discours de la méthode*, p. 60.

[2] Descartes, *Réponses aux secondes objections*, p. 175.

[3] Clairaut, *Eléments de géométrie*, preface, p. 12.

form of the theorem had been stated by Girard and Descartes in the seventeenth century, and many proofs had been proposed. Among the existing proofs in Bolzano's day were ones by d'Alembert, Euler, Lagrange, and Laplace. Gauss proposed three proofs of this theorem in 1799. Many of the existing proofs of the fundamental theorem of algebra used in an essential way the fact that a polynomial over the integers with odd degree must have a real root. That it will have a positive and a negative value is easy to establish. Between these it must be shown that there is a zero. The most common procedure is to use geometrical intuition – the graph must cross the axis. Bolzano criticized this proof because geometrical intuition is used to establish a result in pure mathematics. He added that in science, proofs must not only be supported by evidence, but must expose the objective foundation of the truth being proved. Geometrical intuition is, perhaps, evident, but it is not a proof. Bolzano also rejected proofs based on movement and time which proceed by analogy with moving bodies. To him, such movements were examples, but did not justify the theorem.

A theorem must be fundamentally true, and what makes a proposition fundamentally true is its relation to other propositions. For the existence of a real root of a polynomial over the integers with odd degree, the fundamental idea is that of "continuous function." Bolzano defined a continuous function, showed that polynomials are continuous, and then stated and proved what is now known as the Bolzano-Weierstrass theorem. He introduced the limit of a series and, before Cauchy, introduced the idea of Cauchy convergence.

This break with the concept of proof as convincing evidence was partly due to Bolzano, and partly due to new matters concerning nineteenth century mathematicians. Bolzano's philosophy was opposed to that of Kant, who believed that mathematics only had existence in the mind. Bolzano believed that mathematics and mathematical propositions had an ontological existence; what constituted the existence of a proposition was its relation to other propositions. Bolzano rejected geometric considerations upon which his predecessors had leaned, and this attitude became more widespread. In addition, in the eighteenth century there had been much progress using Cartesian methods and the infinitesimal calculus, which gave algebra, analysis and arithmetic primacy over geometry. And there were problems in geometry arising from the famous parallel axiom of Euclid.

The Break in the Nineteenth Century

The construction of non-Euclidean geometries deepened the rupture between proof as deductive logic and proof as the accumulation of evidence. Some early attempts to deal with Euclid's parallel axiom are interesting. In his *Euclides ab omni Naevo Vindicatus* [Euclid Vindicated from all Faults], published in 1733, Saccheri attempted to prove Euclid's parallel axiom from the other axioms. He started with the quadrilateral, $ABCD$, of the Arab mathematician Omar Khayyam, in which A and B are right angles and $AC = BD$. It is easy to prove that angles C and D are equal. Euclid's axiom is equivalent to the assertion that C and D are right angles. Saccheri considered the two possible alternatives:

- C and D are obtuse angles (the hypothesis of the obtuse angle)
- C and D are acute angles (the hypothesis of the acute angle)

With the first hypothesis Saccheri easily deduced what he believed was a contradiction. (He tacitly assumed the infinitude of a straight line.) With the second hypothesis Saccheri proved many interesting theorems without obtaining any contradiction. He continued until he reached a geometrical situation where the straight lines became too twisted and he wrote, "The hypothesis of the acute angle is repugnant to the nature of the straight line." Between the results of the deductive reasoning and the evidence of the geometrical situation, Sacherri chose the evidence.

Later, however, Bolyai and Lobachevsky chose the deductive reasoning. They constructed non-Euclidean geometries by replacing Euclid's parallel axiom, and the view of mathematicians changed. Each of the three hypotheses – the hypotheses of right angles, obtuse angles, or acute angles – led to non-contradictory propositions. In this way the meaning of "axiom" changed. An axiom no longer had to be based on evidence, but became a free hypothesis.

In his *Grundlagen der Geometrie* [Foundations of Geometry] published in 1899, Hilbert didn't assume that we had any intuition about points, lines, or surfaces. These were only different classes of objects. All we have to know are the relations, given by axioms, between these different classes. Einstein wrote that axioms are free creations of the human mind, and the terms points, lines, etc., in axiomatic geometry must be understood only as schematic concepts devoid of content.

For Legendre, a proposition was true when it produced evidence leading to what he considered a proof. For the formalist school, a proposition is true if it does not contradict the axiomatic system. The idea of contradiction also occurs in Greek geometry, but it has an altogether different sense. For the Ancients, contradiction occurred within a social act and was used to convince another. Today, contradiction occurs within a system of mathematical propositions and it is used to produce mathematical results.

The formalist view influenced the introduction of "modern mathematics" into French school programs. Now the spirit of new syllabi is not so formalist, and the question of evidence arises again. Today, in the study of learning and of proof, can we afford not to include an epistemological reflection on the meaning of proof, and on the role of evidence and of contradiction?

DIDACTIC CONSIDERATIONS

The point here is that the meaning of proof has changed in history. Teachers today should become aware of the fact that the notion of proof is not an absolute. The following observations are offered.

1.   To immediately associate the idea of proof with that of deductive reasoning is not self-evident.
2.   To peremptorily affirm that to prove is to convince is a dogmatic way of approaching the question of the sense of proof.

We have already indicated certain epistemological objections to these concepts. There are also didactic implications. If teachers immediately associate the idea of proof with that of deductive reasoning, if they think that such forms of reasoning will arise "naturally" in their pupils' heads, without supposing that there could be an inherent epistemological difficulty, might they not have set themselves a hopeless expectation?

1. Deductive Reasoning: An Epistemological Obstacle

As we have already noted, the first feeling of proof comes from the very process by which the result is shown. If one asserts that the purpose of proof is a social act designed to convince others, the first feeling can be lost. If you say to a pupil, "This is an explanation, it is not a proof," how will he or she be able to move forward to an understanding of proof (especially if the idea of proof in the mind of the teacher is that of deductive reasoning)? For the teacher, deductive reasoning convinces, whereas the pupil contents himself with being enlightened. Therein lies an epistemological obstacle.

2. Understanding a Proof

The equivalence of "to proof = to convince" poses another problem. If one is convinced by another's proof, must one consent to it, or must that wait until one feels it, forced by the evidence of the truth? Is it an easy matter to make another's proof one's own? Each one of us has had the experience of not having understood a proof that derives from some source, and has had to get hold of it to make it one's own. It is not sufficient to be convinced, one wants to be enlightened. This must not be ignored in learning about proof.

3. Scientific Debate and Socio-Cognitive Conflict

Our point of view leads us to the scientific debate concerning the learning of proof. The work in Geneva, in particular that of Doise and Mugny in *Le développement social de l'intelligence* [The social development of intelligence] certainly agrees with the idea of proof as a social act to convince. Scientific debate in the classroom is certainly effective to initiate students in the ideas of refutation and deduction. Teachers, however, who wish to put socio-cognitive conflict into use would do well to be aware of its limitations – in particular with respect to the learning of proof.

The first difficulty in the practice of scientific debate is that it assumes that everything can come from the mind of the child, including deductive reasoning. The idea that the construction of knowledge should take place "naturally" arising from the social conflicts of children is false. Deductive reasoning is the outcome of the culture of Ancient Greece, Chinese mathematicians were not interested in it, and our pupils are not Greek philosophers – neither are they Chinese mathematicians.

Another limitation concerns the transference of collective results to the individual. The teacher must be aware that collective work in the classroom can be carried out by pupils without there necessarily being any individual perception. This is particularly relevant when learning about proof, for here, as we have already noted, one can be convinced by a proof without having grasped the sense of it, so that one would not be able to reproduce it oneself.

4.   The Motivations for Proof

What is the purpose of proof? Some didactic studies in France suggest that the production of proof is tied to the idea of uncertainty. The proof, for oneself and for others, is produced in order to remove that uncertainty. But there is a problem. If the purpose of proof is to render evident and to make something understood, then it is risky to present it straight away as an "antinomy of the evidence."

To present proof as a way of resolution in the case of uncertainty suggests that the attention of the pupil is turned towards the result. For example, is it true or false that the sum of the angles of a triangle is equal to 180 degrees? Does it matter to the pupil? If the purpose of the teacher is to teach proof, should he or she not find some situations where the learner aims as much, if not more, at the means of obtaining the result as the result itself?

5.   The Teaching of Methods, Methodology, and the Construction of Rationality

Difficulties encountered by pupils in learning about proof suggest that this learning cannot be detached from the construction of mathematical knowledge and the construction of rationality. The example given previously of the even/odd argument can be supplemented by the following given by another teacher. She had worked through a proof concerning a property of triangles with her class. Her pupils were fourteen years old. But one of her pupils was not satisfied. He would not believe that this proof was valid for all triangles, in particular for very thin ones, thinner than the one used in the proof. He had to redo the proof using a thin triangle before agreeing to accept the proof. Teachers often teach the students to cast doubt upon a proof based upon the drawing of one or more particular figures, and then use this idea to show the necessity of a general proof. But what happens if the pupil doubts the validity of a proof based upon a diagram which represents a general case?

The epistemological break that occurred in the seventeenth century leads us to pose the following question. If learning the idea of proof necessitates the existence of rationality on the part of the pupil, what are the didactic virtues of a methodological approach? In *Le rationalisme appliqué*, Bachelard writes, "How, for instance, can we ignore the pedagogical aspect of the enumeration of knowledges advised by Descartes? This methodological inspection ... obliges us to be conscious of our rational identity through the diversification of acquired knowledges. Their order orders us."[1] Rationality is an individual act by which a person becomes "a being of knowledge," whereas the idea of proof as a social act pushes us to research social interactions.

From this point of view, what didactic situations should be put into practice? Since it concerns the construction of methods, we need to start from situations which give rise to the necessity of orderly and methodic procedures. At first, all methods of obtaining a result are acceptable, but the pupil needs to be asked to explain his method. It is this self-reflection which, according to Bachelard, constitutes an action of applied rationalism and allows the pupil to become a "being of knowledge." Piaget claims that pupils as young as seven years old can self-reflect. In

---

[1] Bachelard, *Le rationalisme appliqué*, p. 14.

Le jugement et le raisonnement chez l'enfant he rates highly the ability of the conscious mind to become aware of its consciousness. He writes, "The logical justification of a judgement ... requires reflection and language, in short, an introspection which constructs above the spontaneous thinking a 'thinking of thinking' only able of logical necessity."[1] This remark seems particularly interesting with respect to the learning of proof.

Concerning the learning of proof, one should not ask "How do you prove this result?" but, "How have you obtained this result?". This should be the first step followed by the perfecting of methods of proof arising from confrontations with those produced by the pupils themselves. Then, step by step, one could arrive at a more subtle method of proof which is that of deductive reasoning.

I would like to finish by asking, "Does not the construction of the rationality of the pupil have priority in the teaching of mathematics?"

References

[1]   Archimède, *Oeuvres complètes*, trad. Paul Ver Eecke, Vaillant-Carmannes, Liège, 1960.

[2]   Arnauld, *Noveaux éléments de géométrie* (1667), ed. I.R.E.M. de Dijon, 1982.

[3]   Arnauld et Nicole, *La logique ou l'art de penser* (1674), ed. PUF, Paris, 1965.

[4]   Bachelard, *Le rationalisme appliqué*, PUF, Paris, 1949.

[5]   Barbin, Heuristique et démonstration in mathématiques. La méthode des indivisibles au XVIIème siècle, in *Fragments d'histoire des mathématiques*, n°2, A.P.M.E.P., Paris, 1987.

[6]   Barbin, Les éléments de géométrie de Clairaut: une géométrie problématisée, in *Repères I.R.E.M.*, n°4, juillet 1991.

[7]   Barbin, La démonstration mathématique: histoire, épistémologie et enseignement, in *Actes des 2émes journées Paul Langevin*, ed. Rosmorduc, Université de Brest, to appear.

[8]   Bolzano, Démonstration purement analytique... (1817), trad. Sebestick, *Revue Franç. d'hist. des Sc.*, 1964.

[9]   Clairaut, *Eléments de géométrie* (1765), ed. SILOE, Laval, 1986.

[10]   Descartes, *Les méditations métaphysiques*, PUF, Paris, 1961.

[11]   Descartes, *La géométrie* (1637), Christophe David, Paris, 1705.

---

[1] Piaget, *Le jugement et le raisonnement chez l'enfant*, p. 121.

[12]   Descartes, *Les règles pour la direction de l'esprit*, Vrin, Paris, 1966.

[13]   Descartes, *Le discours de la méthode*, Flammarion, Paris, 1966.

[14]   Doise et Mugny, *Le développement social de l'intelligence*, Interéditions, Paris, 1981.

[15]   Euclid, *Eléments*, trad. Peyrard, Blanchard, Paris, 1966.

[16]   Fowler, D. H., "Ratio in early Greek mathematics," Bulletin of the A.M.S., 1(1979), 807-846.

[17]   Kline, *Mathematical Thought from Ancient to Modern Times*, Oxford University Press, 1972.

[18]   Piaget, *Le jugement et le raisonnement chez l'enfant*, Delachaux et Niestlé, Neufchâtel, 1967.

[19]   Szabo, Greek dialectic and Euclid's axiomatics, in Lakatos, *Problems in the philosophy of mathematics*, North Holland Company, 1972.

[20]   Vernant, *Mythe et pensée chez les Grecs*, Maspero, Paris, 1971.

Some Mathematical Applications from the History of Cryptology for the Mathematics Classroom

Helen Skala
University of Wisconsin/LaCrosse
LaCrosse, WI 54601

In the past two decades cryptology has attracted considerable attention within the mathematical community. To a large extent, this interest is due to the widespread use of computers for the transmission and coding of messages. However, cryptology has been fascinating to mathematicians throughout recorded history. In this paper we will look at a few historical examples which provide interesting mathematical activities for the classroom.

First, lets look at some terminology in this field: plaintext – the message to be put into secret form; ciphertext – the coded message; cryptography – methods of encoding text; cryptanalysis – methods of decoding text; cryptology – includes both cryptography and cryptanalysis; cipher – the substitution of each letter of the alphabet by one or more symbols; code – the substitution of whole words or phrases by other words, symbols or phrases.

Both ciphers and codes seem to have been used as far back as Babylonian times. One of the oldest Babylonian ciphers dates back to 1500 B.C. and appears on a tablet found on the site of ancient Seleucia on the banks of the Tigris. It contains the earliest known coded formula for the making of glazes for pottery. However, the world owes its first instructional text on cryptology to the Greeks. It appeared in a military treatise, "On the Defense of Fortified Places" by Aeneas the Tactician, in which a number of cryptographic systems were described. One example is a substitution cipher in which letters are replaced by combinations of dots.

Substitution ciphers in which letters are replaced by different symbols have been common throughout the ages. An esoteric cipher devised by Charlemagne, which probably caused his generals many a sleepless night trying to memorize it, was the following:

A B C D E F G H I K L M N O P Q R S T U X Y Z

A simpler cipher, whose origin is unknown, is called the pigpen cipher, and is much easier to remember:

53

A letter is enciphered by writing the lines and dots surrounding it. The word "love", for example, would be written as:

$$< \sqcup \cdot \sqcap \cdot \square$$

One need not, however, invent new symbols to use a substitution cipher. Perhaps the best known substitution cipher is one attributed to Julius Caesar, in which letters are encoded by translating them a fixed number of places in the alphabetic sequence. For example, A is replaced by F, B by G, C by H, etc., so that the message "help" would be enciphered as MJQU.

Another early example of a cipher system, which is not a substitution cipher, consists of a device called a "scytale" and was used as early as the fifth century B.C. by the Spartans. A strip of parchment is wrapped around a staff and the message written on it. When the parchment is unwrapped, the letters are disconnected. Only if one has a staff of the same dimensions, can the message be decoded. The scytale is the earliest example of a "transposition" cipher, one in which the letters of the plaintext are scrambled. Let's look more closely now at these two basic ciphers, the substitution and transposition ciphers, to see what mathematics is involved.

Caesar's cipher is a substitution cipher known as a "shift transformation". If we number the letters in the alphabet with 0 for A, 1 for B, 2 for C, etc., we can describe Caesar's cipher by the equation, $c \equiv p + k \pmod{26}$, where c denotes the cipher letter, p the plaintext and k the amount of shift. In order to decipher the message, we must apply the inverse of this transformation, which would be: $p \equiv c - k \pmod{26}$. An easy way to perform this addition or subtraction is to make two strips having the sequence 0 through 25 twice. By moving one strip right or left relative to the other we find the sum or difference of two numbers modulo 26.

Instead of adding a constant, we could multiply by a constant - for example, $c \equiv 3p \pmod{26}$. Multiplication by 3 is a one-to-one transformation. This is a necessary condition for such a transformation to work for encryption. Students may discover that multiplication by k is one-to-one provided k and 26 are relatively prime. Moreover, the inverse would also be a one-to-one multiplication transformation. In order to find the multiplication factor, we must solve the equation $km \equiv 1 \pmod{26}$ for m.

If we combine these two transformations, addition and multiplication, we end up with what is called an affine transformation: $c \equiv ap + b \pmod{26}$. Its inverse is $p \equiv e(c - b) \pmod{26}$ where $ae \equiv 1 \pmod{26}$. An interesting question is whether any letters are "fixed" by an affine transformation. That is, does the equation $x \equiv ax + b \pmod{26}$ have any solutions? It turns out that this equation has a solution if and only if the gcd of $a - 1$ and b divides 26.

The Greek "scytale," as mentioned earlier, is an example of a transposition cipher – that is, the letters of the plaintext are scrambled. Some interesting geometrical questions with regard to the actual scytale device can be posed. For example, how many letters may be written along the staff? The answer, of course, depends on the width of the parchment and the length and diameter of the staff. However, the answer cannot be obtained by a simple division because the parchment must be wrapped on a slant. Some trigonometry is needed.

Let g denote the length of the staff, r the radius of the staff, and w the width of the parchment. Imagine we have wrapped the parchment around the staff, then removed the staff and finally flattened the parchment. The parchment will then be folded in a zigzag fashion as in the following figure.

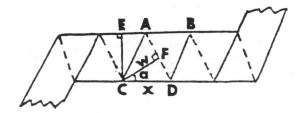

Draw CE perpendicular to AB and CF perpendicular to AD. Hence CF $= w$. Also CE is half the circumference of the staff, so CE $= \pi r$. Let $\angle DCF = a$ and CD $= x$. Then since AD $=$ AC, $\angle ACD = \angle FDC = \pi/2 - a$ and since $\angle ECD = \pi/2$, $\angle ECA = a$. Hence in right triangle CFD, $\cos a = w/x$ and in right triangle ECA, $\tan a = (x/2)/(\pi r)$. Multiplying these equations, we obtain $(\tan a)(\cos a) = \sin a = w/(2\pi r)$. Finally,

$$x = \frac{w}{\cos a} = \frac{w}{\cos\left(\sin^{-1}\left(\frac{w}{2\pi r}\right)\right)} = \frac{w}{\sqrt{1 - \left(\frac{w}{2\pi r}\right)^2}} = \frac{2\pi r w}{\sqrt{4\pi^2 r^2 - w^2}}.$$

And the number n of letters that can be written across will be

$$n = \left[\frac{g}{x} + 1\right] = \left[\frac{g\sqrt{4\pi^2 r^2 - w^2}}{2\pi r w} + 1\right]$$

where we make the convention that a letter will be written on the partial end sections of the parchment and [ ] denotes the greatest integer.

Another question we may ask is: given the length and the radius of the staff and the width of the parchment, how long should the parchment be in order to completely cover the staff? In order to solve this problem we note that when the parchment covers the entire staff, the area of the parchment equals (approximately) the surface area of the staff. That is, if r and g again denote the radius and length of the staff and if w and f denote the width and length of the parchment, then $wf = 2\pi rg$ or $f = 2\pi rg/w$.

Now suppose a message has k letters in it and we wish to write the message so that it covers the whole staff. How can we determine the dimensions of the parchment and staff? In order to answer this question, we must also determine how many "lines" m of text can be written around the staff. The value m will depend on the height h of the letters. If r, g, w, f, and n are as above, then we have the following equations.

$$m = [2\pi r/h] \text{ and } k = mn = \left[\frac{2\pi r}{h}\right] \times \left[\frac{g\sqrt{4\pi^2 r^2 - w^2}}{2\pi r w} + 1\right]$$

Since the parchment must completely cover the staff, we also have $f = 2\pi rg/w$. (Note that w should be less than $2\pi r$.) Hence if k, the number of letters in the message, is given, any three of the variables r, g, w, h can be specified and the remaining one determined by the above equations.

Most early cryptology was confined to cryptography, that is, creating methods of enciphering. There were no general methods known to decipher a message. The first people to address the question of cryptanalysis were the Arabs. Early Arabic knowledge of cryptology was described in an encyclopedia which was completed in 1412 by the scholar Qalqashandi. Much of this work on cryptology can be traced back to hundreds of years prior to the completion of the encyclopedia. The section on cryptology in Qalqashandi's encyclopedia opens with a remarkable exposition, the first written description of a cryptanalytic method to decipher a simple substitution cipher. Qalqashandi begins like this. "When you want to solve a message which has been written in code, count the letters and, for each letter, set down the total. When you have determined which letter occurs most frequently assume that it is alif, the next most frequent should be lan ...." Here was the first description of frequency analysis, and perhaps even the first application of inferential statistics.

Frequency analysis successfully "cracked" all simple substitution ciphers. For each language, one can make a frequency table of the percentage of occurrences of the letters in its alphabet. In English a typical table is:

A	B	C	D	E	F	G	H	I	J	K	L	M	N
7.3	0.9	3.0	4.4	13.0	2.8	1.6	3.5	7.4	0.2	0.3	3.5	2.5	7.8

O	P	Q	R	S	T	U	V	W	X	Y	Z
7.4	2.7	0.3	7.7	6.3	9.3	2.7	1.3	1.6	0.5	1.9	0.1

These frequencies may vary somewhat. Correspondences dealing with war matters, for example, are likely to have a different frequency table than those dealing with culinary techniques.

Since frequency analysis provided a nearly foolproof way to decode simple substitution ciphers, new methods of cryptography were being developed by the end of the fifteenth century. One of the most clever cryptologists of this time was Matteo Argentis, who was employed by the Vatican in the early sixteenth century. He invented the following cipher to foil frequency analysis techniques.

A	B	C	D	E	F	G	H	I	L	M	N	O	P
1	86	02	20	62	22	06	60	3	24	26	84	9	66
				82									

Q	R	S	T	U	V	Z	ET	CON	NON	CHE	nulls
68	28	42	80	04	46	88	08	64	00	44	5,7
				40	48						

The message "Lasciate ogni speranza" could be encoded as: 24154 20275 31805 82906 84342 76662 28184 87817 (Since 5's and 7's are nulls, they are ignored.) It is interesting to speculate why these particular numbers were chosen. Can you pose some conjectures? The order of frequency for the Italian language is typically

E A I O N L R T S C D P U M V G H F B Q Z.

Another method developed at this time was that of polyalphabetic substitution, invented by a Benedictine monk Johannes Trithemius in the early 1500's. He created what he called a "tabula recta" and which today is usually referred to as a Vigenere Table.

To use Trithemius' table, the first letter of the plaintext message is enciphered by using one row of the table, the second letter by another row and so on. Which rows are used is usually determined by some keyword. For example, if the keyword is JUSTINE, then the first letter is enciphered using the J-row, the second using the U-row, etc., cyclically. The message "Stay back" would be encoded as BNSRJNGT using the keyword JUSTINE.

Without knowledge of the keyword, the polyalphabetic substitution cipher was considered to be unbreakable until the 19th century, when some general methods of solution were described. The most important item needed to solve any polyalphabetic cipher is the length of the keyword. If one knows the key length k, then the solution of the polyalphabetic cipher is reduced to solving k Caesar ciphers.

An important approach to discovering the key length was described by a Prussian military officer named F. W. Kasiski in 1863. He instructed the cryptologist to look for pairs of repeated sequences of three or more letters and to assume that the repetition of these letters occurred because the same plaintext letters were enciphered by the same sequence of alphabets. If one counts the number of letters between a repeated occurrence, then the key length must divide that interval length. If one finds a few repeated pairs, one can usually narrow down the key length to a small number of possibilities.

```
    a b c d e f g h i j k l m n o p q r s t u v w x y z
A   A B C D E F G H I J K L M N O P Q R S T U V W X Y Z
B   B C D E F G H I J K L M N O P Q R S T U V W X Y Z A
C   C D E F G H I J K L M N O P Q R S T U V W X Y Z A B
D   D E F G H I J K L M N O P Q R S T U V W X Y Z A B C
E   E F G H I J K L M N O P Q R S T U V W X Y Z A B C D
F   F G H I J K L M N O P Q R S T U V W X Y Z A B C D E
G   G H I J K L M N O P Q R S T U V W X Y Z A B C D E F
H   H I J K L M N O P Q R S T U V W X Y Z A B C D E F G
I   I J K L M N O P Q R S T U V W X Y Z A B C D E F G H
J   J K L M N O P Q R S T U V W X Y Z A B C D E F G H I
K   K L M N O P Q R S T U V W X Y Z A B C D E F G H I J
L   L M N O P Q R S T U V W X Y Z A B C D E F G H I J K
M   M N O P Q R S T U V W X Y Z A B C D E F G H I J K L
N   N O P Q R S T U V W X Y Z A B C D E F G H I J K L M
O   O P Q R S T U V W X Y Z A B C D E F G H I J K L M N
P   P Q R S T U V W X Y Z A B C D E F G H I J K L M N O
Q   Q R S T U V W X Y Z A B C D E F G H I J K L M N O P
R   R S T U V W X Y Z A B C D E F G H I J K L M N O P Q
S   S T U V W X Y Z A B C D E F G H I J K L M N O P Q R
T   T U V W X Y Z A B C D E F G H I J K L M N O P Q R S
U   U V W X Y Z A B C D E F G H I J K L M N O P Q R S T
V   V W X Y Z A B C D E F G H I J K L M N O P Q R S T U
W   W X Y Z A B C D E F G H I J K L M N O P Q R S T U V
X   X Y Z A B C D E F G H I J K L M N O P Q R S T U V W
Y   Y Z A B C D E F G H I J K L M N O P Q R S T U V W X
Z   Z A B C D E F G H I J K L M N O P Q R S T U V W X Y
```

The modern Vigenère tableau

Another important approach to discovering the length of the keyword was suggested by the American cryptologist William Friedman in 1920. His idea was to compute some statistic for an enciphered message and then compare that value with the value one would expect in the monalphabetic case. This statistic he called the "index of coincidence" and defined it as

$$\text{I.C.} = \sum_{n=A}^{Z} f_n \cdot f_{n-1}$$

where f_n is the frequency of the nth letter of the alphabet in the enciphered message. If a monalphabetic substitution is used then the I.C.'s for some languages are

English: .066, French: .076, German: .076, Portuguese: .079, Russian: .053, Spanish: .078.

Polyalphabetic ciphers tend to equalize the frequencies of letters so that f_n is approximately 1/26 for each letter (in English). For this "flat" distribution, the I.C. = .038. Hence polyalphabetic ciphers will have an I.C. closer to .038. The cryptologist can use the I.C. to discover the length of the keyword by calculating the I.C. of sets of letters consisting of every ith letter for $i = 2,3,...$. These sets are called components of the message. For example, if the length of the keyword were 5, then if we calculate the I.C. of the component containing the first letter, the sixth letter, the eleventh letter, etc., we should obtain approximately .066 for English since all of these letters were enciphered with the same alphabet. Similarly if we calculate the I.C. of the component containing the second letter, the seventh letter, the twelfth letter, etc., we should likewise obtain .066 approximately. There will be five such components, each of whose I.C. should be about .066. For example, consider the following coded message.

```
KRAHM KPTVN UFUZQ TFLML VRKAK YRTEK ITURQ TPRKV QSBHU KPAKY SKGEQ RGJUF NEOTN
SZHVG IKKTW EGURT IOFTV OAGTO BINKP AKSTT FZTXT KNXGT FAGRQ TUZGG GGEBP KELCT
SXTOY ICZUX PIQHT OYIGX FBPKZ VGOZA KYYTP TQKKT AMJKD PQXPL JGPMJ KDXDK QGCTK
MJOZZ QXPBP GDBNE TNOGZ TDUGM JOYWQ ZNMNK ELKYT HWRPA CBQWG YYBUY QWJOY YTUYM
JKBKG SULGY
```

The I.C. of this message is .047. Hence we suspect a polyalphabetic encipherment. The next step is to calculate the I.C. of the message assuming an i-letter keyword for $i = 2,3,....$ The I.C.'s for various key lengths are:

Assumed key length	I.C. of components					Average I.C.
2	.050	.057				.053
3	.053	.047	.043			.047
4	.064	.080	.069			.067
5	.047	.035	.040	.056	.053	.046

A key length of 4 seems to give reasonable I.C.'s for this cryptogram.

The next step is to determine the keyword. Assuming key length of 4, we first write down the letters of the first component, consisting of the first letter and every fourth letter thereafter. Each of these letters was encoded by the same Caesar shift. Friedman suggested that to determine this shift, one should once again use the table of relative letter frequencies.

The cipher text for the first component is KMVUFVKE... The corresponding plain component, depending on what the Caesar shift was, would be:

Amount of shift	Possible plaintext	Sum of log-percentages
0	KMVUFVKE...	35.9
1	JLUTEUJD...	69.6
2	IKTSDTIC...	122.4
3	HJSRCSHB...	50.6
etc.		

Using the relative frequency table, we would like to determine the likelihood of each of the above possible sequences occurring in a plaintext message. Although individual letter frequencies are not independent, we might use the product

$$\text{freq}(K) \cdot \text{freq}(M) \cdot \text{freq}(V) \cdot \text{freq}(U) \cdots$$

as a heuristic measure. However this product becomes very small quickly, so a related measure called the "sum of log-percentages", which gives values that are neither very small nor exceedingly large, can be used instead.

$$\log(100 \cdot \text{prob}(K)) + \log(100 \cdot \text{prob}(M)) + \cdots$$

The larger this measure is, the more likely it is that the letters are part of a plaintext message. In the above, a shift of 2 gives the highest value and hence we assume the keyword starts with C. To find the second letter of the keyword we work with the second component. Continuing with this method, we find the keyword is "CGMT".

The deciphered message is "I looked at him steadfastly. His face was leanly composed. His grey eye dimly calm. Not a wrinkle of agitation rippled him. Had there been the least uneasiness, anger, impatience or impertinence in his manner, in other words, had there been anything ordinarily human about him, doubtless I should have violently dismissed him from the premises." (from Herman Melville's, *Bartleby the Scrivener*).

The Greek "scytale", as mentioned previously, is a different type of cipher, called a transposition cipher, in which the letters of the plaintext are scrambled. An effect similar to the scytale can be achieved by using a "row to column" encipherment. That is, write the plaintext row by row in a matrix of n columns and then write the ciphertext by following the columns. For example "send more money" could be written using 3 columns as

$$
\begin{array}{ccc}
s & e & n \\
d & m & o \\
r & e & m \\
o & n & e \\
y & q & u
\end{array}
$$

and the ciphertext would be SDROYEMENQNOMEU. Note that if the message does not completely fill the matrix, add extra characters to complete the last row. The I.C. is again useful in determining if a message is a transposition cipher. Its I.C. would be the same as the I.C. for plaintext.

Instead of writing the cipher text by following columns we might follow some other route, such as along diagonals. Or, to generalize even more, why confine ourselves to rectangular matrices? Any geometric shape could be specified, along with a particular route to follow for encipherment.

Another matrix based cipher system suggested by Lester Hill in the early 1900's uses matrix multiplication. First write the plaintext in matrix form with numbers substituted for letters. For example, A = 0, B = 1, etc. Ciphertext is obtained by multiplying the plaintext matrix by a given invertible matrix. For example, the message "come now" would first be converted to numbers "2, 14, 12, 4, 13, 14, 22"; using a 2 by 4 matrix form, we write the matrix

$$
A = \begin{bmatrix} 2 & 14 & 12 & 4 \\ 13 & 14 & 22 & 3 \end{bmatrix}
$$

where numbers are inserted randomly to fill out the matrix. Using the invertible matrix

$$
M = \begin{bmatrix} 3 & 2 \\ 7 & 5 \end{bmatrix} \text{ with inverse}
$$

$$
M^{-1} = \begin{bmatrix} 5 & -2 \\ -7 & 3 \end{bmatrix} \text{ we encode the message by multiplying } M \times A \text{ to get}
$$

$$
M \times A = \begin{bmatrix} 32 & 70 & 80 & 18 \\ 79 & 168 & 194 & 43 \end{bmatrix} = \begin{bmatrix} 6 & 18 & 2 & 18 \\ 1 & 12 & 12 & 17 \end{bmatrix} (\text{mod } 26)
$$

and the enciphered message is GSCSBMMR. To decode the message, write the coded message as a matrix and multiply it on the left by M^{-1}.

This encipherment may be made more complicated by breaking the plaintext into more than one matrix and using different enciphering matrices to encode the separate matrices; or by computing linear combinations of these separate matrices to garble the message even more.

This discussion would not be complete without mentioning the now-famous public key cryptosystems which are practically unbreakable at this time. These systems were developed to enable users to send messages over public communication channels without fear of decipherment. They are based on theorems from elementary number theory.

Each user of the system chooses two large prime numbers p and q each of which has more than 100 digits. Since there exist computer algorithms to generate large primes very quickly, this task is not as complicated as it might seem. Let $n = pq$ and choose a large number e relatively prime to $m = (p-1)(q-1)$. Since e is relatively prime to m, there exists a number d such that $ed \equiv 1 \pmod{m}$. Again, there are programs that can find such e's and that can calculate d quickly. Each user makes his chosen numbers n and e public information but keeps d, p and q secret. Because p and q have more that 100 digits, it is, at this time, virtually impossible to find the factors of n. Hence p, q and d will remain secret.

Now suppose someone wishes to send a message to a person whose public keys n and e are given. The message is first changed to a numeric form by the correspondence A = 01, B = 02, etc. and partitioned into blocks of reasonable length. For example, "It is all Greek to me" could be written as 092009 190112 120718 050511 201513 05. The sender then computes $c \equiv b^e \pmod{n}$ for each block b and transmits these numbers c to the recipient. The recipient, in turn, computes $c^d \pmod{n}$ for each c to obtain the original message. (Anyone else using the system will also have access to these numbers but, without knowing the number d, will find the message impossible to decipher.)

To see how this method works, let's use some smaller numbers. Let $p = 3$ and $q = 11$. Then $n = pq = 33$ and $m = (p-1)(q-1) = 20$. Choose $e = 3$ which is relatively prime to 20. Then since $3 \cdot 7 \equiv 1 \pmod{20}$, $d = 7$. Suppose someone wants to send you the message "help". Using two digits per block we convert this message to 08 05 12 16. The sender computes 8^3, 5^3, 12^3, and 16^3 modulo 33, which are 17, 26, 12, and 4. The receiver then computes 17^7, 26^7, 12^7, and 4^7 modulo 33 which are 8, 5, 12, and 16, the original message.

A program which does much of the cryptanalysis described in this paper is available from the author. Please send a diskette formatted for IBM-compatibles (either $5\frac{1}{4}''$ or $3\frac{1}{2}''$) and a self-addressed, stamped envelope.

Some Sample Problems for Students

1. Use Charlemagne's substitution cipher to encipher the following message.
   "Send more troops"

2. Cipher the plaintext "Strike at dawn" using the following transformation.
   $$c \equiv 7p - 3 \pmod{26}$$

3. Suppose you know that the following message was encoded by an affine transformation $c \equiv ap + b \pmod{26}$: YHTQF SCUFD SBULX IOLBF ALYZT IDSCL YCSDO FZYCU FAFMF ODITF YCDKV SBICK SBSCF TX . Find a and b by solving the pair of simultaneous equations obtained by matching the two most frequently occurring letters in the coded message with e and t, the two most frequently occurring letters in English plaintext. [Solve $5 \equiv 4a + b$ and $2 \equiv 19a + b$ or $18 \equiv 19a + b$. Number theory has played an important role in the development of cryptosystems.]

4. The following is a typical order of frequency for letters in the English language.
   E T O A N I R S H D L C W U M F Y G P B V K X Q J Z
   Decode the following message, which is a monalphabetic cryptogram.

   YAMAB QYLSM QVOEE TMKME MLQMZ YBDBK KBUPO SAY

5. Use the matrix $M = \begin{bmatrix} 2 & 0 & 1 \\ 3 & 1 & 2 \\ 1 & 0 & 1 \end{bmatrix}$ to encipher "congratulations."

   Use M^{-1} to decipher

   "TWPAQLAAFVMLWSNFTLEECZWN."

REFERENCES

[1] Camp, Kane, "Secret Codes with Matrices", *Mathematics Teacher* (Dec 1985), 676-678.

[2] Dror, Asael, "Secret Codes", *Byte Magazine* (June 1989), 267-270.

[3] Feltman, James, "Cryptics and Statistics", *Mathematics Teacher* (Mar 1979), 189-191.

[4] Friedman, William, *Elements of Cryptanalysis*, Government Printing Office, Washington DC, 1924.

[5] Gass, Frederick, "Solving a Jules Verne Cryptogram", *Mathematics Magazine* (Feb 1986), 3-11.

[6] Gardner, Martin, *Codes, Ciphers and Secret Writing*, Simon and Schuster, New York, 1972.

[7] Grady, M. Tim and Doug Brumbaugh, "From Greeks to Today: Cipher Trees and Computer Cryptography", *J. of Computers in Mathematics and Science Teaching* (Summer 1988), 56-59.

[8] Kahn, David, *The Codebreakers*, Macmillan, New York, 1967.

[9] Kuenzi, N.J. and Bob Prielipp, *Cryptarithms and other Arithmetical Pastimes, School Science and Mathematics Assoc.*, Topics for Teachers Series, No. 1, 1979.

[10] Laffin, John, *Codes and Ciphers–Secret Writing through the Ages, Abelard-Schuman*, New York, 1961.

[11] Lucians, Dennis and Gordon Prichett, "Cryptology: From Caesar Ciphers to Public-Key Cryptosystems", *The College Math Journal* (Jan. 1987), 2-17.

[12] Mann, Barbara, "Cryptography with Matrices", *Pentagon* (Fall 1981), 3-11.

[13] Menceley, Merrill, "Decoding Messages", *Mathematics Teacher* (Nov. 1961), 629 632.

[14] Peck, Lyman, *Secret Codes, Remainder Arithmetic and Matrices*, NCTM, 1961.

[15] Schwartz, Charles, "A New Graphical Method for Encryption of Computer Data", *Cryptologia* (Jan 1991), 43-46.

[16] Snow, Joanne, "An Application of Number Theory to Cryptology", *Mathematics Teacher* (Jan 1989), 18-26.

[17] Wolfe, James, *Secret Writing–The Craft of the Cryptographer*, McGraw Hill, New York, 1970.

Reflections of a Problems Editor

Clayton W. Dodge
University of Maine
Orono, ME 04469

What good is problem solving? Problems are the heart of life. Life itself is solving problems. So here is a sample of life itself.

Many educators are reluctant to use calculators and computers in the classroom, and teachers have always denied technical advances. When I was in grammar school, for example, we had to use pencil for all our writing because we were too young to use pens. In junior high school we were not allowed to use fountain pens because we "had to learn to use pen and ink properly." When I reached high school we were required to use fountain pens and could not use those new-fangled ball-point pens.

I was given a dollar slide rule and taught how to multiply and divide with it when I was in grammar school, but I could not take it to school; it would have been confiscated as contraband. One afternoon in 1942 shortly after World War II had started, when I was in the fifth grade, I wandered into my father's shoe store and noticed all the men sitting around the store writing. When I asked my father about it, he said that the Office of Price Administration (OPA) required the wholesale and retail prices of all goods in the store and the percentage of markup based on cost and also on selling price, calculated to three digits. So he had one man gathering the figures and the other seven salesmen doing the pencil-and-paper long divisions. I told him I could do those divisions easily on my slide rule. So that Saturday the eight salesmen gathered figures and I did the divisions, a system more than *fifty times* as efficient as their earlier method. Some months later my father showed me a letter he had just received from the OPA commending him for being the first store in that city of 60,000 people to turn in his figures and the only one to do it correctly the first time!

The moral of that story is not that I did anything remarkable, but that I had a tool and knew how to use it. If we try to deny modern technology, we are burying our heads in the sand. Indeed, new developments challenge *us* to think and they give us the tremendous opportunity to teach real understanding, to work with *all* the tools that are in our mathematical tool kit when faced with a problem.

Let us turn to a selection of problems, mostly in geometry, that I have run across over the years, problems that are interesting to me for various reasons.

One asks, "Have I seen a problem similar to this one? Can I use the same method on the current problem? Can I generalize the method I used on another problem?" For example, the *checkerboard problem* is well known. A standard 8 by 8 checkerboard can be covered with thirty-two 2 by 1 *dominoes*. If a square is removed from each of two opposite corners of the board, can the remaining 62 squares be covered by 31 dominoes (see Figure 1)? The secret to an easy solution lies in the fact that a checkerboard has squares of two colors, and that a domino will cover exactly one square of each color. Since the two squares that are removed have the same color, as shown in Figure 2, there remain 30 squares of that color and 32 of the other color, so 31 dominoes will not cover the truncated board.

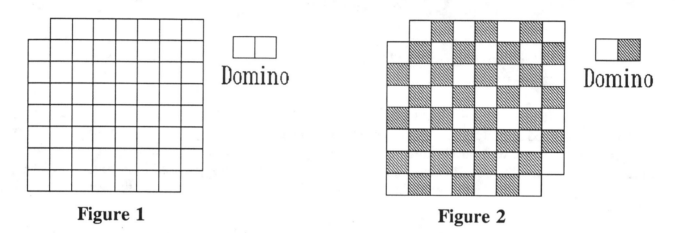

Figure 1   **Figure 2**

Now let us look at *The Hexagon Problem* [1], proposed by Sidney Penner in the *Pi Mu Epsilon Journal*. In Figure 3, delete one of the "*A*" hexagons surrounding the center hexagon in this parallelogram array of $(2n + 1)^2$ hexagons. Can the remaining hexagons be covered with non-overlapping trominoes?

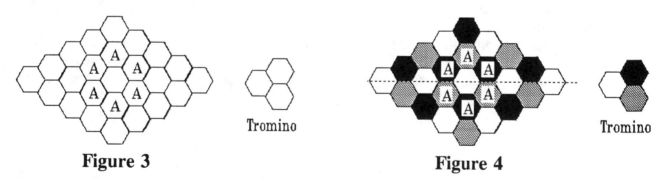

Figure 3   **Figure 4**

As in the checkerboard problem, let us consider coloring the hexagons. Since we are covering with trominoes, let us use a tromino coloring with three colors. This is shown in Figure 4 using gray shading and solid black. No matter where a tromino is placed, it covers three hexagons of different colors. The tromino may have to be rotated or flipped over for the colors to match. The parallelogram array now has a horizontal line of symmetry such that, for each hexagon that is gray, there is a mirror-image hexagon that is black, and vice versa. Hence there are just the same number of black hexagons as gray, and each tromino will cover just one of each. When we remove an "*A*" hexagon, we upset that balance, so that trominoes can no longer cover the figure.

Let us leave geometry briefly for an unsolved problem of cloudy origins that appeared shortly before World War II. We call it the *Collatz Problem*. Although I had heard of it earlier, I first saw it in print in *Eureka* [2]. Let *f* be the function that takes a positive integer *n* to $n/2$ if n is even or to $3n + 1$ if n is odd. Prove or disprove that any positive integer can be reduced to 1 by successively applying *f* to it. H. S. M. Coxeter has offered \$50 for a proof or \$100 for a disproof, Paul Erdös offered \$500, and B. Thwaites £1000 for a solution.

In 1985 Lagarias [3] published an excellent article about the Collatz problem. Many curious features have been observed about this function. For example, by applying *f* to 12 and to 13 we get the sequences

$$12 \to 6 \to 3 \to 10 \to 5 \to 16 \to 8 \to 4 \to 2 \to 1$$

and

$$13 \to 40 \to 20 \to 10 \to 5 \to 16 \to 8 \to 4 \to 2 \to 1.$$

Each takes 9 steps. Why should two adjacent numbers take the same numbers of steps? It is easy to show that all pairs of the forms $8n + 4$ and $8n + 5$ merge after the third application of *f*. Now 26 takes 10 steps, while 28, 29, and 30 each take 18 steps. Curiously, 27 takes 111 steps! The interested reader may wish to verify these facts. Problems such as this one provide excellent opportunities for school teachers to give their students cleverly disguised arithmetic drill. It is easy to write a computer program to calculate successive values of *f* and to count the number of steps to reach 1. The patterns one sees in the collected data are most intriguing. Inspired by the Collatz problem, I wrote an article [4] that told how the early programmable calculators could be forced to simulate conditional branching and calculate $n/2$ if n is even or $3n + 1$ if n is odd.

I cannot resist telling of a beautiful geometry problem that is discussed in Howard Eves' delightful book *Mathematical Circles Revisited* [5]. In 1943 Howard Eves ran the *Butterfly Theorem*, proposed by Sol Mitchell, in the Elementary Problem Department of the *American Mathematical Monthly* [6]: Let *O* be the midpoint of a chord of a given circle, as shown in Figure 5, and let *TU* and *VW* be any two other chords through *O*. Let *TW* and *UV* cut the given chord in points *E* and *F*. Then *O* is the midpoint of *EF* also.

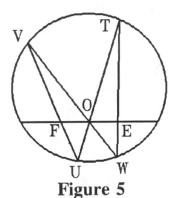

Figure 5

Solution IV, a delightful generalization of the theorem to any conic section, was by Emory P. Starke, for many years the Editor of the *American Mathematical Monthly* problem departments. A few years later Howard Eves received a letter from Starke, asking for his help because he had recently come across a delightful little theorem but could not find a proof of it. Howard recognized it as the butterfly theorem and simply sent Starke a postcard with just the reference to the journal page numbers on which his own solution appeared.

Four years ago I gave essentially this same lecture as the J. Sutherland Frame Lecture following the Pi Mu Epsilon banquet at the combined summer mathematical meetings in Salt Lake City, Utah. John Wetzel then told me of two of his favorite problems in geometry, which I later took the liberty to publish [7].

Call a plane arc *special* if it has length 1 and lies on one side of the line through its end points. Prove that any special arc can be contained in an isosceles right triangle of hypotenuse 1.

Proof: Let *PQ* be a special arc as shown in Figure 6, and construct the smallest isosceles right triangle *ABC* with hypotenuse *AB* on line *PQ* that contains arc *PQ*. Of course, that triangle is the unique circumscribed isosceles right triangle with hypotenuse on line *PQ* (see Figure 7). Let arc *PQ* touch the legs of the triangle at *R* and *S*. Now reflect arc *PR* in line *AC* and arc *SQ* in line *BC*, obtaining arcs *RP'* and *SQ'*. Then arc *PRSQ* equals arc *P'RSQ'* in length and lies between lines *AP'* and *BQ'* which are distance *AB* apart. The arc length is 1 and is not less than *AB*, proving the theorem. Wetzel concludes "Pause and reflect!"

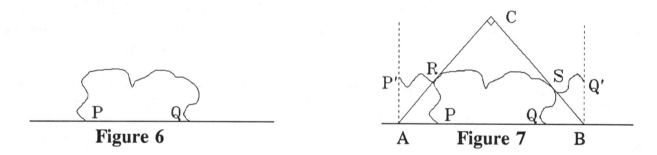

Figure 6   A   **Figure 7**   B

Wetzel's second problem, part of an article he later wrote [8], involves the figure for Napoleon's theorem, which states: If equilateral triangles *BCP*, *CAQ*, and *ABR* are constructed outwardly on the sides of a given triangle *ABC*, then the centroids of the equilateral triangles form a triangle whose centroid coincides with the centroid of triangle *ABC*. Furthermore, the lines *AP*, *BQ*, and *CR* concur. In 1868 Lemoine asked whether one can reconstruct triangle *ABC* if one knows *P*, *Q*, and *R*. Kiepert soon provided the affirmative answer: Let *X*, *Y*, and *Z* be the third vertices of outward equilateral triangles constructed on the sides of triangle *PQR*. Then *A*, *B*, and *C* are the midpoints of the segments *PX*, *QY*, and *RZ*, a delightful bit of geometry (see Figure 8).

Kiepert's argument is not especially pretty, so Wetzel proposed the following transformational proof. Let α denote the composition of the three 60° rotations about point *P*, point *R*, and point *Q* in that order. Then α is a halfturn. The rotations in α take point *C* to *B*, then to *A*, and back to *C*, so *C* is a fixed point for α. Therefore α is a halfturn about *C*. Now the rotations in α map point Z to Q, then to X, and finally to R. Since then α, a halfturn about C, maps Z to R, then C is the midpoint of RZ. Similarly, A and B are the midpoints of PX and QY.

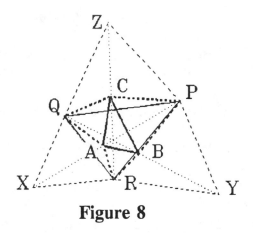

Figure 8

Some thirty years ago I assisted Howard Eves with the Elementary Problem Department of the *American Mathematical Monthly*. A retired mathematician sent in a proposal to prove that the apothem OS of a regular hexagon is equal to the side of the regular heptagon (7-sided polygon) inscribed in the same circle, as shown in Figure 9. If we let the radius of the circumcircle be 1, then it is easy trigonometry to show that the apothem is given by $OS = \cos(\pi/6) \approx 0.8660254$ whereas the side of the heptagon is $2\sin(\pi/7) \approx 0.86777$, differing from OS by only about 0.2%, a remarkably close approximation.

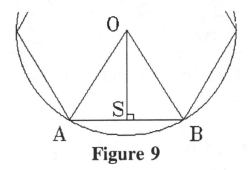

Figure 9

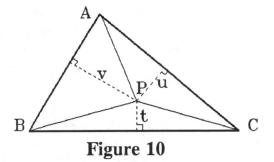

Figure 10

I carefully wrote up the trigonometry and sent it back to the proposer, rejecting the proposal. In a few days I received a most scathing letter. The proposer complained that I had disproved his theorem by using the trigonometric functions, which are "only approximations," but he had done the construction on his garage floor with a 10-foot sheet of paper and 5-foot compasses, so he *knew* it was correct! I didn't answer that letter, but for the next month I looked very carefully for any suspicious characters as I left my office each day.

The Erdös-Mordell inequality was posed in the *Monthly* in 1935 by Paul Erdös [9] and solved by L. J. Mordell two years later [10]. A delightful second solution appeared in 1957 by D. K. Kazarinoff [11]. The problem states that for any triangle *ABC* and any point *P* within or on the triangle, the sum of the distances from P to the vertices of the triangle is greater than or equal to twice the sum of the distances from P to the sides of the triangle (see Figure 10). That is,

$$PA + PB + PC \geq 2(t + u + v).$$

In 1974 Huseyin Demir proposed the *Extended Erdös-Mordell Inequality* [12], that the inequality above holds for all points P anywhere in the plane provided that a perpendicular distance to a certain side shall be taken negative if P and the vertex opposite that side are separated by the side. That is, t, for example, will be taken negative if A and P lie on opposite sides of BC.

I was assigned to referee the solutions for that problem. There were a total of four solutions from four different countries throughout the world. Curiously, each solver had used Kazarinoff's solution, which involves multiplying three inequalities by the distances t, u, and v, and then adding. Unfortunately, each solver overlooked the fact that at least one of t, u, and v is negative in the extension, reversing the corresponding inequality. The three inequalities cannot then be added, and the proof fails. Leon Bankoff and I corresponded for two years trying unsuccessfully to patch the proof. I was convinced of the truth of the theorem and returned to it from time to time. During a two-week March school break some ten years later I set aside all else and finally found a solution which was published in 1984 [13]. My proof is far from elegant, so a nice, clean proof is still desired.

The enticing feature in this problem is that it is easy to dispose of the case where P lies outside the circumcircle of the given triangle. The inequality is either obvious or else it reduces to a case where P lies inside the circumcircle but outside the triangle. Curiously, Kazarinoff claimed without demonstration that his proof holds for any point P lying within the circumcircle of the given triangle, provided the distances were given the appropriate sign convention when the point lay outside the triangle. If Kazarinoff's claim is true, then the elegant proof exists already, but there is no record to substantiate his claim and it is not at all obvious that his published proof does hold for all points inside the circumcircle.

Another problem of interest I refereed, albeit not in geometry, was entitled *Sets of Numbers with Equal Sum and Product* [14]. It stated: For every $k > 1$ there is a set of k positive integers whose sum and product are equal. For $k = 2, 3, 4, 6, 24$, the set is unique. Is it unique for any other k?

The solution that holds for every k is $\{k, 2, 1, ..., 1\}$. The published solution by E. P. Starke found second solutions for all other $k < 1000$ except 114, 174, and 444. The problem was interesting to me because it was the first problem to which I applied a computer. By examining a printout of computer second solutions, I found patterns for certain values of k. In fact, general patterns were found for all $k > 6$ except multiples of 6 that end in the digit 4. My computer search showed no other exceptional k to 15,000, and more recently, to 60,000.

The *arbelos* is formed by three mutually tangent semicircles as shown in Figure 11. It is the figure formed by the semicircles (O), (O_1), and (O_2), whose centers are the points O, O_1, and O_2. The line ACB contains their diameters. It is also called the *shoemaker's knife* because arc AC is shaped like the handle and arc CB like the blade of a cobbler's knife. Now erect a perpendicular *CD* to *AB* at *C*. Then the *twin circles of Archimedes* are the two circles (W_1) and (W_2), each tangent to line *CD* and to two of the semicircles. It is easy to show that the twin circles are congruent with radius equal to half the harmonic mean of the radii of circles (O_1) and (O_2).

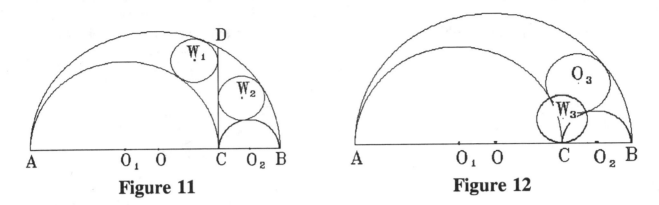

Figure 11   **Figure 12**

Leon Bankoff wrote an article entitled The Twin Circles of Archimedes; Are they Really Twins? (*Mathematics Magazine*, 1974, pp. 214-8) [15]. In it, he finds another circle congruent to the twin circles. So his answer to the question is "No, they are two of triplets." Last year, when I was asked to give a lecture to students, I stole his title, and attempted to come up with some other circles. After a few weeks of unsuccessful puzzling, I called Leon and he told me of a fourth circle he had found. After the phone call I located a letter I had written to Leon in 1975 telling him of two more sibling circles. It reminded me of Starke's experience with the butterfly theorem. Anyway, that discovery shook loose my cobwebs and I managed to find several more sibling circles to present in the lecture.

The *triplet circle* of Bankoff is shown in Figure 12. Circle (O_3) is drawn inside the arbelos tangent to each of the first three circles. Then the triplet circle is the circle (W_3) through point *C* (tangent to line *AB*) and the two points of tangency of circle (O_3) with circles (O_1) and (O_2). The proofs of these results are most easily obtained by inversion. Although the twin circle theorem is easily proved with high school geometry, such a proof for the triplet circle is messy.

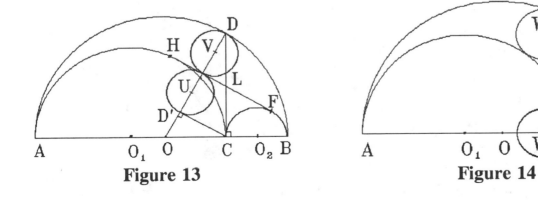

Figure 13   **Figure 14**

Bankoff's fourth circle, which is also one of my two circles, is shown in Figure 13 with my other circle. Here we have drawn the external tangent line *FH* to the two smaller circles of the arbelos. Bankoff's fourth circle (*V*) is the largest circle tangent to that line and to the outer circle (*O*). Curiously, it is tangent to the outer circle at *D*.   If you draw radius *OD* and drop a perpendicular *CD'* from *C* to *OD*, it is easy to show that *DD'* is the harmonic mean of the diameters of circles (*O₁*) and (*O₂*), so two Archimedean circles (*U*) and (*V*) fit on that segment. These are the circles I found, although I did not independently discover that *FH* was their common tangent.

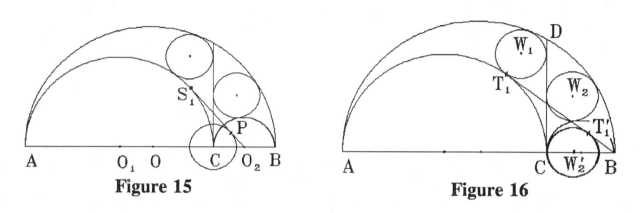

Figure 15   **Figure 16**

Searching for other circles, I projected circles (*W₁*) and (*W₂*) onto the diameter *AB*, obtaining (*W₁'*) and (*W₂'*) of Figure 14. The circle (*C*) on *W₁'W₂'* as diameter is also an Archimedean circle, bringing the total now to 8 congruent circles. If these circles were merely projections, their existence would indeed be dull. It turns out that they have other interesting properties to justify the prominence. Figure 15 shows that circle (*C*) is the circle centered at *C* and tangent to the line *O₂S₁* that is drawn from *O₂* tangent to circle (*O₁*). Figure 16 shows that circle (*W₂'*) is tangent at *T₁'* to the line *BT₁* drawn from point *B* tangent to circle *O₁*. Point *T₁* is also the point of tangency of the circles (*O₁*) and (*W₁*). By symmetry the tangents to circle (*O₂*) from *O₁* and from *A* are tangent to the Archimedean circles (*C*) and (*W₁'*).

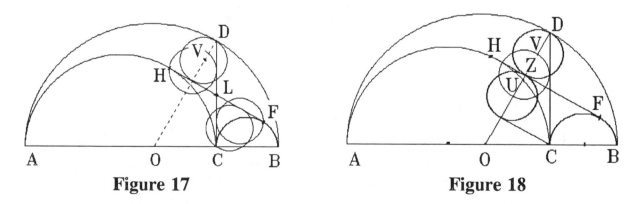

Figure 17   **Figure 18**

My final observation came from Figure 13 and circle (*V*). Because *CD = FH* and *L* is the midpoint of *CD* and also of *FH*, and because (*V*) is the circle tangent to *FH* at a point diametrically opposite point *D*, there are three more circles symmetrically located in that figure (see Figure 17).

I ended my lecture with the challenge to the students to try to find other circles congruent to these eleven. Afterwards one of the students remarked that there is one more obvious circle, shown in Figure 18, the circle (Z) on *UV* as diameter.

We have shown that there are at least 12 congruent circles in this delightfully elegant arbelos. Are there more such circles? Thus I leave you with a question, a challenge, in hopes that you will be enticed to delve into problems.

REFERENCES

[1]   Penner, S., Problem 434, *Pi Mu Epsilon Journal*, 7 (Fall 1979) 72-73.

[2]   Williams, K. S., Problem 133, *Eureka* (now *Crux Mathematicorum*), 2 (1976) 144-150.

[3]   Lagarias, J. C., The $3x + 1$ Problem and its Generalizations, *The American Mathematical Monthly*, 92 (1985) 3-23.

[4]   Dodge, C. W., Overcoming Calculator Indecision, *Crux Mathematicorum* 6 (1980) 70-71.

[5]   Eves, H. W., *Mathematical Circles Revisited*, Prindle, Weber & Schmidt, Boston, (1971) Item 245°.

[6]   Mitchell, S., Elementary Problem E 571, *American Mathematical Monthly* 50 (1943) 326 and 51 (1944) 91-92.

[7]   Wetzel, J. E., Problems 759 and 760, *Pi Mu Epsilon Journal* 9 (1991) 316.

[8]   Wetzel, J. E., Converses of Napoleon's Theorem, *American Mathematical Monthly* 99 (1992) 339-351.

[9]   Erdös, P., Problem 3740, *American Mathematical Monthly* 42 (1935) 396.

[10]   Mordell, L. J., Solution to Problem 3740, *American Mathematical Monthly* 44 (1937) 252.

[11]   Kazarinoff, D. K., A Simple Proof of the Erdös-Mordell Inequality for Triangles, *Michigan Mathematical Journal*, 4 (1957) 97-98.

[12]   Demir, H., Elementary Problem E 2462, *American Mathematical Monthly*, 81 (1974) 281.

[13]   Dodge, C. W., The Extended Erdös-Mordell Inequality, *Crux Mathematicorum* 10 (1984) 274-281.

[14]   Simmons, G. J. and D. B. Rawlinson, Elementary Problem E 2262, *American Mathematical Monthly*, 77 (1970) 1008.

[15]   Bankoff, L., The Twin Circles of Archimedes; Are they Really Twins?, *Mathematics Magazine* 47 (1974) 214-218.

Mathematical Creativity in Problem Solving and Problem Proposing II

Murray S. Klamkin
University of Alberta
Edmonton, Alberta
Canada   T6G 2G1

Dedicated to Howard Eves

I. Introduction

No doubt all of us here are very pleased to be at this conference in honor of Howard Eves on his 80th birthday. My first communications with Howard started back in 1948, when I was a beginning instructor at what was known then as the Polytechnic Institute of Brooklyn and I had begun to submit a rather large number of problem proposals to the American Mathematical Monthly. Howard was the elementary problem editor and maintained that position for 23 years. He and Emory Starke, the advanced problem editor, also for very many years, were both very dedicated and knowledgeable editors as well as being dedicated and knowledgeable problemists. Since probably at that time, in retrospect, most of my proposals were not particularly good, they just got filed away. Unfortunately for me, this included one favorite one. This was to determine whether or not one could remove one square from a chess board and cover the deleted board with 21 3×1 trominoes. This was no loss to problem solvers, since several years later, Solomon Golomb's book [1] on polyominoes came out which included this problem plus many more goodies. My next aquaintance with Howard came in reading his nice paper on Philo's Line [2] and I will be discussing some of the problems raised there subsequently. My last spoken communications to him, excluding this conference, had taken place quite some years ago when I gave some invited talks at the University of Maine. I had spoken there about problem solving and I will now be talking on a similar theme.

In all my talks on problem proposing and/or problem solving, I always highly recommend four of George Pólya's books [4-6]. Even though there have been other books on the subject, to me, the Pólya books are far superior.

In Hilbert's address [7-9] to the I.C.M. in Paris in 1900 on "Mathematical Problems" which is remembered chiefly because of its list of 23 important unsolved problems, he also included quite a bit on the importance of problems to mathematics and the important roles of *generalization* and *specialization*. This was taken up later and expanded upon by Pólya. Since I think these remarks of Hilbert's are neglected, especially when compared to his list of unsolved problems, I have quoted most of them in the preface of my recent book on applied problems [10]. However, since there are mathematicians who tend to denigrate "problemists," I now quote one other paragraph from Hilbert's address:

> "The deep significance of certain problems for the advance of mathematical science in general and the important role which they play in the work of an individual investigator are not to be denied. As long as a branch of science offers an abundance of problems, so long is it alive; a lack of problems foreshadows extinction or the cessation of independent development. Just as every human undertaking pursues certain objects, so also mathematical research requires its problems. It is by the solutions of problems that the investigator tests the temper of his steel; he finds new methods and new outlooks, and gains a wider freer horizon."

I make no claim about the importance of the problems that I will treat. Nevertheless, small solved and unsolved problems lead to larger solved and unsolved problems which in turn lead to important mathematical results. As an example, consider the mathematics developed in attempts to prove Fermat's "Last Theorem" which even if true is not particularly important. A metaphor attributed to the late Allen Shields is particularly appropriate here, "A mathematical problem is a 'jackpot' which increases in value as we throw our quarters into it."

In a previous address with the same title (part I, [3]), I had noted some factors necessary to be creative, e.g., one should be observant, persistent; and then gave quite a few examples. Here I'll let the development of the problems themselves implicitly indicate the aspects of creativity involved. That one can see this may be indicative of the relatively low level of the difficulty of the problems. Unfortunately, we do not see in the writings of many top mathematicians how they were led to their results. A notable exception is Euler. No doubt, some mathematicians are not concerned with including these aspects; and no doubt a number of editors would have deleted them if included. Also, there are cases when the mathematician is unaware of how he came to his results. As an illustration of this, K. Knopp once noted that Gauss described his discovery of an important number-theoretical result in 1805 in the following way:

> "But all the brooding, all the searching was for nothing; finally, a few days ago I succeeded. But not by long searches but by the sheer grace of God, I may say, like lightning strikes, the riddle was solved; I myself would not be able to find the connection between what I knew previously, with what I used for my last attempt and with what finally succeeded."

Particulary apropos here is the book of Hadamard on the psychology of mathematical invention [11].

Before continuing with the examples, let me briefly outline "horizontal" problem solving as opposed to "vertical" problem solving [12]. By the latter, I mean the usual way where one is studying some branch of mathematics, e.g., linear algebra, and solves problems in succession by using the theorems developed. This could involve different methods. By the former method, I mean one which uses some general problem pattern, e.g., symmetry, continuity, and uses this as a key idea to unlock the solution of problems from different branches of mathematics. I had used both methods in my ten years of coaching the USA Mathematical Olympiad Teams and these teams have had an excellent record. However, their success may have been due more to their innate ability then to the coaching. A number of these patterns are treated in the afore mentioned books of Pólya. Here is an expanded listing that I have given out in my problem solving courses.

A. FIRST CONSIDER
1. SPECIAL CASES.
2. GENERALIZATIONS.
3. ANALOGIES (MATHEMATICAL or PHYSICAL).
4. RELATED PROBLEMS

B. PATTERNS
1. PSYCHOLOGICAL BLOCKS.
2. SYMMETRY.
3. CONTINUITY.
4. SIMILARITY, HOMOTHETICITY, DIMENSIONALITY.
5. REDUCTIO AD ABSURDUM (INDIRECT PROOF).
6. MATHEMATICAL INDUCTION.
7. PIGEON HOLE PRINCIPLE.
8. RELAXATION (REDUCTION OF DIMENSIONALITY).
9. GENERATING FUNCTIONS.
10. LEVEL CURVES AND SURFACES.
11. LINEARITY.
12. CONVEXITY.
13. INEQUALITIES.
14. WORKING BACKWARDS.
15. INGENIOUS SUBSTITUTIONS.
16. SOLVING DIFFERENTLY AND COMPARING.
17. EQUIVALENCE CLASSES.
18. TRANSFORMATIONS.
    a. TRANSLATION.
    b. ROTATION.
    c. REFLECTION.
    d. INVERSION.
    e. AFFINE.
    f. CONICAL.
    g. RECIPROCATION.
    h. MISCELLANEOUS.

Some of these patterns will be illustrated briefly in subsequent problems. To cover them all in depth would take, at least, a mini-course.

II. An Optimization Problem

My first problem is an optimization one that I had seen in a calculus book during my visit to New Caledonia College in Prince George, B.C. It concerned a farmer who wished to fence in a maximum area of land in the form of a trapezoid with a given length of fencing. One of the bases was not to be fenced since it was along a straight river. The hint given with the problem was to introduce the altitude of the trapezoid and treat the problem by Lagrange multipliers with four variables and one constraint. Let me digress for a moment. Many years ago I had made a survey of maxima and minima problems in a large number of our calculus texts and found that almost all of them could be done much more simply by other methods, in particular, by the A.M.-G.M. inequality in two and three variables. I can see using a number of such problems for some relatively easy illustrations of Lagrange multipliers, but this should have been pointed out and also a number of real calculus problems should have been included. For non-calculus methods of handling extremum problems, see [12-14].

Returning to our problem (see Fig. 1), we want to maximize $[ABCD]$ given that $BC||AD$ and $x+y+z=p$ (here and subsequently, $[F]$ will denote the area or volume of a figure F).

Since many problems can be simplified by symmetrization, we reflect $ABCD$ (Patterns #2,18c) across AD to form a hexagon of twice the area and twice the perimeter p. We now invoke the isoperimetric inequality for polygons [15], i.e., of all $n-$gons with given perimeter, the regular one has the maximum area and dually of all $n-$gons with given area, the regular one has a minimum perimeter. Consequently, $[ABCD]$ is a maximum when $x=y=z$ and the base angles A and $D=60°$.

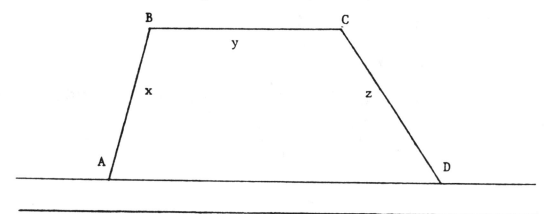

Figure 1

A relatively easy way to be creative in problem proposing is to start with a given problem and solution and try to extend it or generalize it. I. Kaplansky [16] in an interview quotes a joke due to R. Boas which is a play on an old car waxing advertising slogan, "Be wise, Simonize." His motto was "Be wise, Generalize." The roles of generalization emphasized by Hilbert and Polya have already been noted (Good problem strategies had been given much earlier by the Bernoullis).

For a related problem, we add the extra constraint that $\angle A$ is also given. Here we now have to maximize

$$x\{x \cos A + 2p - 2x - 2z + \sqrt{z^2 - x^2 \sin^2 A}\,\}$$

which is a function of the two variables x and z. We can reduce the dimensionality of the problem (Pattern #8) by first holding x fixed so that we first want the minimum value of

$$2z - \sqrt{z^2 - x^2 \sin^2 A}\,.$$

By ordinary calculus methods, this occurs when $3z^2 = 4x^2 \sin^2 A$. We can also obtain this result by the A.M.-G.M. inequality, (Pattern #13) by first letting $z = (x \sin A)(e^\theta + e^{-\theta})/2$ which reduces the expression to $(x \sin A)(e^\theta + 3e^{-\theta})/2$. Substituting for z back in (1), we now want the maximum of

$$x\{2p - 4x \sin^2 (A + 60°)/2\}.$$

Again by the A.M.-G.M. inequality, $x = p/4 \sin^2 (A + 60°)/2$. Letting $\angle A = 60°$, we recapture our previous results.

We can continue in the same vein by also requiring that $\angle D$ be given. Here the problem is simpler and reduces to maximizing

$$h\{2p - h(2 \csc A + 2 \csc B - \cot A - \cot B)\}$$

where h is the altitude of the trapezoid. By the A.M.-G.M.,

$$h = p/(2 \csc A + 2 \csc B - \cot A - \cot B).$$

Again we recapture our previous results by letting $\angle A = \angle D = 60°$.

III. Bisecting Arcs

The next problem is a very nice one in Pólya [6, p. 186]; I managed to catch D.J. Newman on it as well as other mathematicians (see introductory remarks on using geometry in [17]). A bisecting arc is one which bisects the area of a given region. First, I asked what is the shortest bisecting arc of a circle. Usually, the fast reply is that it is a diameter. Secondly, I asked what is the shortest bisecting arc of a square. Again, a usual fast reply is that it is an altitude through the center. Finally, I asked what is the shortest bisecting arc of an equilateral triangle. By this time, Newman had suspected that I was setting him up (and I was) and almost was going to say the angle bisector. But he hesitated and said let me consider a chord parallel to the base and since this turns out to be shorter than an angle bisector, he gave this as his answer. Unfortunately for him, the correct answer is a $60°$ arc of a circle.

Incidentally, it is a very rare occasion when I can catch Newman on problems such as these. To solve this problem, we again use reflection (Pattern #18c). This time we successively reflect the equilateral triangle about successive bases to form a regular hexagon as in Fig. 2. Finally, we invoke the isoperimetric theorem for circles [15]; of all plane figures with given area the circle has the least perimeter and dually of all plane figures with given perimeter the circle has the maximum area.

The problem for the circle is a variation of a 1946 Putnam problem [18] and the more general solution given in Pólya [6, p. 272] for centro-symmetric regions can be simplified. A simple solution for circles is given in [18] (this latter book did not appear until 1980).

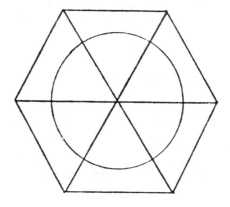

Figure 2

Here's a solution for centro-symmetric regions: Let A and B be the end points of the bisecting arc and O the center (Fig. 3).

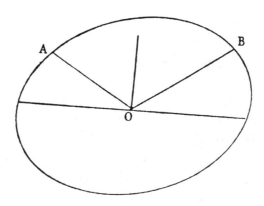

Figure 3

Draw the chord through O which is perpendicular to the angle bisector to $\angle AOB$. It now follows by reflection that $AO + OB$ is the minimum length of a curve from A, B to the chord through O. Hence the shortest bisecting arc must be greater than or equal to $AO + OB$ and so the shortest bisecting arc is the shortest chord through O. This proof also applied to the USAMO problem #3, 1976 [19] which was to show that if two boundary points of a unit ball are joined by a curve contained within the ball and having length less than 2, then the curve is contained entirely within some hemisphere of the ball.

Pólya also noted that the shortest bisector of any region is a straight segment or an arc of a circle. This, however, is at variance with the results in an apparently little known paper of N. Wiener [20] in which he proves that the shortest bisecting arc of a region will be either an arc of a finite or infinite circle, or will be a chain of such arcs such that two successive arcs only meet on the boundary of the area. Wiener also comments that "it is almost self-evident that the shortest line to divide a convex area in a given ratio is a single arc of a circle (*finite or infinite*), but this I have not been able to prove." As far as I know, this is still unproved. For other related problems, see [21].

Two other problems which fall out quickly by symmetry are the following:

(i) If in E^n we have $n + 1$ concurrent unit vectors at a point P that are equally inclined to each other, then their sum vanishes [22]. One can give a proof using linear algebra but a simpler proof follows by indirect proof and symmetry (Pattern #2,5). We assume that the sum which is unique does not vanish. Then by the group of motions which take the configuration into itself, there would be many different sums and this is impossible. The common angle between any pair of the vectors turns out to be $\cos^{-1}(-1/n)$ and is obtained simply by expanding out the square of the sum of the $n + 1$ vectors. Also, in E^n there cannot exist more than $n + 1$ vectors equally inclined to each other. Geometrically, the end points of the $n + 1$ vectors are the vertices of a regular simplex whose centroid, circumcenter and incenter are all at P.

(ii) It is immediate that the gravitational attraction on a test particle at the center of a uniform centro-symmetric body is zero. The same result applies for the test particle at the center of a uniform tetrahedron. However, centro-symmetry does not apply here but the previous proof in (i) does.

IV. Philo's Line and Variations

Our next set of problems is Philo's line and variations thereof. Referring to Figure 4, we want the extreme values of $F(x, y, u, v)$ for different functions F given the point P and $\angle A$. Some of these are the following:

    A. $\min (x + y)$; Philo's line,

    B. $\min [ABC]$,

    C. $\min xy$; 1976 Putnam problem,

    D. $\max(1/x + 1/y)$; 1979 USAMO problem,

    E. $\min$ perimeter (ABC).

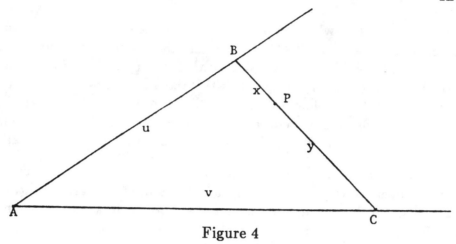

Figure 4

These problems have also occurred in other competitions as well as repeatedly in various journal problem sections. I'll now sketch out some of the solutions. Other nice alternative ones are given in a recommended article by Howard Eves [23]. Also for **A**, see [24].

A. This version when $\angle A = 90°$ is given in quite a number of calculus books. We take a rectangular coordinate system and let the point P be (h, k) and the line through it be $x/a + y/b = 1$. Then we want to minimize $a^2 + b^2$ subject to $h/a + k/b = 1$. We get an immediate solution (both necessary and sufficient conditions) by applying Hölder's inequality, i.e.,

$$(a^2 + b^2)^{1/3}(h/a + k/b)^{2/3} \geq h^{2/3} + k^{2/3}$$

and with equality if and only if $a^3/h = b^3/k$. Hence,

$$\min (a^2 + b^2)^{1/2} = (h^{2/3} + k^{2/3})^{3/2}.$$

Generalizations of this problem are gotten by considering 3-dimensional analogues. We take a point (h, k, ℓ) in the first octant of a rectangular coordinate system and pass a plane through it intersecting the coordinate planes in a triangle. We now can consider finding the minimum area of the triangle or the minimum perimeter. The area case leads to minimizing

$$b^2 c^2 + c^2 a^2 + a^2 b^2 \quad \text{subject to} \quad h/a + k/b + \ell/c = 1.$$

So far I have not succeeded with this problem using inequalities or even Lagrange multipliers. Not surprisingly, the minimum perimeter problem has also been intractible. However, if we change it to minimizing the sum of the squares of the sides, then there is an easy solution. Here we want to minimize $a^2 + b^2 + c^2$ subject to the same constraint as before. Also as before, the solution follows by an immediate application of Hölder's inequality, i.e.,

$$(a^2 + b^2 + c^2)^{1/3}(h/a + k/b + \ell/c)^{2/3} \geq (h^{2/3} + k^{2/3} + \ell^{2/3}).$$

This also extends immediately to n-dimensions [14].

For general angles A, the solution of the initial problem is rather messy [23-24] except for the case when P is on the angle bisector. Then by symmetry, BPC is perpendicular to the angle bisector. This very simple result has quite a number of extensions which are not as immediate [25].

B. A geometric solution follows by doubling AP and completing a parallelogram as in Figure 5.

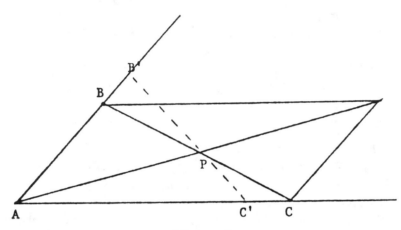

Figure 5

The diagonal BPC cuts off the minimum area. To see this, perturb BPC to $B'P'C$ and it follows easily that $[B'PB] > [C'PC]$.

C. Here we just draw BPC perpendicular to the angle bisector of A. Now just consider the circle (Figure 6) tangent to the sides of the angle at points B and C and use the power of a point theorem for circles (Pattern #10), i.e., the product of the segments of a chord through a given point is constant.

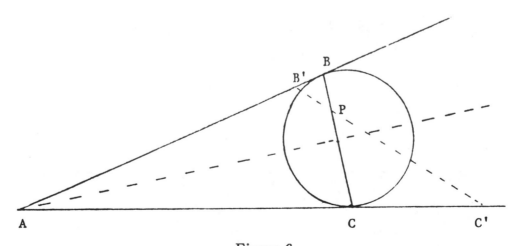

Figure 6

D. I leave this as an exercise. A solution appears in [19, p. 66].

E. This is a problem that I came across when I was in high school and it took me about two weeks working off and on before I solved it. It is a marvelous problem to indicate the use of "Patterns." The difficulty with this problem is that the key geometric theorem that unlocks the solution is unexpected. Very many years later, I was in charge of writing a programmed text in geometry which included a large problem solving section for the Minnemast Project under Paul Rosenbloom. After quite some time I managed to figure out some motivation for the solution by using a number of the listed patterns. My solution here will be much more abbreviated than in the text, since there when a pattern was employed, quite a few different illustrative examples were used to demonstrate the pattern so they would "stick."

It is to be noted that an analytic geometric-calculus approach is straightforward (again see introductory remarks on using geometry in [17]). This leads to minimizing the not too simple expression

$$x\{(k-mx)\sqrt{1+m^2}+k-hm+m\sqrt{(k-mx)^2+(x-h)^2}\,\}/(k-mx)$$

by differentiating with respect to x. I leave it as an exercise to find the zero of the derivative.

In solving extrema problems it is often very useful to consider level curves (Pattern #10) as in C. Here we want to determine a chord(s) through P which cuts off a triangle(s) of fixed perimeter $2s$. Since, at this stage, this related problem also seems difficult we relax (Pattern #8) the problem by not requiring the chords to pass through P. Then two such triangles are easy to construct (specialization) and these are two degenerate triangles whose length of sides are $s, s, 0$. Although the other fixed perimeter triangles are not as yet easy to construct, we sketch some of them as in Fig. 7.

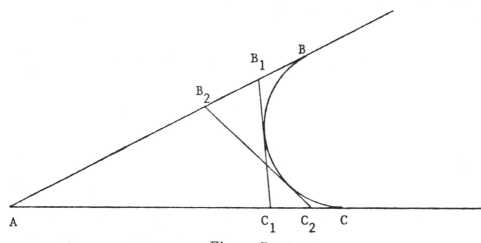

Figure 7

If we knew the curve which is tangent to the lines $AB, AC, B_1C_1, B_2C_2, \ldots$ our problem would be substantially over. Such a curve is called an envelope and at this stage we make a guess as to what it is. Although one may think of a parabolic arc, let us consider a simpler one, the circular arc. Then this should bring to mind the theorem that tangents from an external point to a circle are equal. This by the way is the key theorem for solution and it is not surprising that one wouldn't have focused on it earlier. It now follows from the tangent property (Fig. 8) that $C_1 D = C_1 C$, $B_1 D = B_1 B$ and hence the perimeter of $AB_1 C_1 = AB + AC = 2s$.

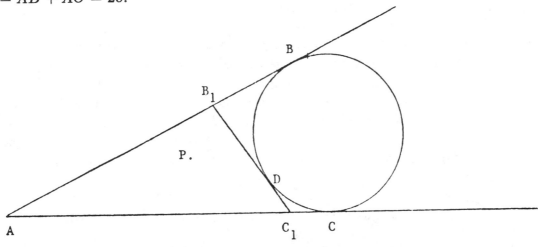

Figure 8

We now "unrelax" the problem by choosing a chord $B'C'$ which contains P and is tangent to the circle. Finally, we return to our original problem. Since the perimeter is $2s = 2AB$, we want to determine a circle tangent to the sides of the angle such that its tangent points are as close as possible to the vertex A. But this must be done under the restriction that we can draw a tangent line from P to the circle. Thus, the desired circle is the one which is tangent to the sides of the given angle and passes through P. This is one of the classic constructions due to Apollonius and, for completeness, I give a solution again using patterns. First by symmetry, the center of the circle is on the angle bisector of A. We now use homotheticity (Pattern #4) and construct any tangent circle as in Figure 9.

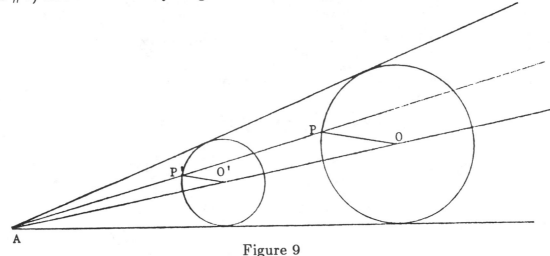

Figure 9

The desired circle and the arbitrary circle just constructed are homothetic with A as the center of homotheticity. We now draw the line AP and then draw the line through P parallel to $P'O'$ to determine O.

Naturally, in using the various patterns you have to have some relevant knowledge. If you did not know the tangent properties of a circle or the properties of homothetic figures, the patterns used would not have been useful.

There is still more to be said about this problem by using working backwards (Pattern #14) and solving differently and comparing (Pattern #16). We will simplify our results by letting $\angle A = 90°$. Let P have coordinates (h, k) in a rectangular coordinate system. Then the desired circle to give the minimum perimeter triangle has the equation $(x-r)^2 + (y-r)^2 = r^2$ where r satisfies $(h-r)^2 + (k-r)^2 = r^2$ and so $r = h+k+\sqrt{2hk}$. Now let $x/a + y/b = 1$ be a variable line through P so that $h/a + k/b = 1$. The perimeter of the triangle formed by this line and the coordinate axes is $a + b + \sqrt{a^2 + b^2}$. This yields the inequality

$$a + b + \sqrt{a^2 + b^2} \geq 2(h + k + \sqrt{2hk}) \tag{1}$$

where $h/a + k/b = 1$. The equality condition occurs when the line $x/a + y/b = 1$ is tangent to the circle. This can be shown to occur when

$$a = r\sqrt{2hk}/(k + \sqrt{2hk}) \quad \text{and} \quad b = r\sqrt{2hk}/(h + \sqrt{2hk}). \tag{2}$$

Finally, we can express the inequality in a, b in the homogeneous form

$$(hb + ka)(a + b + \sqrt{a^2 + b^2}) \geq 2ab(h + k + \sqrt{2hk}). \tag{3}$$

Inequality (3) does not look easy to prove directly. To do so, we first make the natural substitutions $h = x^2$, $k = y^2$ and $a/b = (\sin\theta)/\cos\theta$, which give

$$(x^2 \cos\theta + y^2 \sin\theta)(1 + \cos\theta + \sin\theta) \geq 2(\sin\theta \cos\theta)(x^2 + xy\sqrt{2} + y^2) \tag{4}$$

for all real x and all θ in $[0, \pi/2]$ and with equality if and only if

$$\tan\theta = (y^2 + xy\sqrt{2})/(x^2 + xy\sqrt{2}).$$

Inequality (4) is now apparent since it is equivalent to

$$\{x\sqrt{\cos\theta}\sqrt{1 + \cos\theta - \sin\theta} - y\sqrt{\sin\theta}\sqrt{1 + \sin\theta - \cos\theta}\}^2 \geq 0.$$

V. A Swedish Olympiad Problem

I now give some extensions of a Swedish Olympiad problem [26] which was to find the maximum area rectangle whose pair of opposite sides are chords to two concentric circles of radii a and b. Referring to Figure 10, we want to maximize.

$$[ABCD] = 2x\{\sqrt{a^2 - x^2} + \sqrt{b^2 - x^2}\ \}.$$

Even though a calculus solution is easy, here is an elementary solution by Dan Sokolowsky. Since $[ABCD] = 4[BOA]$, we get the maximum area when $BO \perp OA$ or when

$$\sqrt{a^2 + b^2} = \sqrt{a^2 - x^2} + \sqrt{b^2 - x^2}\ .$$

Then, by squaring twice, $x = ab/\sqrt{a^2 + b^2}$ and $\max[ABCD] = 2ab$.

We now look for possible extensions. One immediate one is to maximize

$$x(\sqrt{a^2 - x^2} + \sqrt{b^2 - x^2} + \sqrt{c^2 - x^2}\ )$$

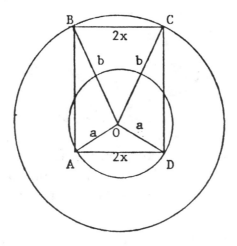

Figure 10

which turns out to be intractible (just consider the complicated derivative equation). Another possible extension is to replace the rectangle by a trapezoid (Figure 11).

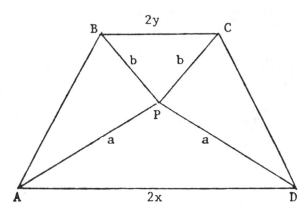

Figure 11

Here we want to maximize the two-variable expression

$$(x + y)\{ \sqrt{a^2 - x^2} + \sqrt{b^2 - y^2} \}.$$

This could make a good problem for multivariate calculus. However, the geometric solution is quite easy. The area of each of the four triangles is maximized if the four angles at P are right angles. So, equivalently, the diagonals of the trapezoid are orthogonal and of length $a + b$. The maximum area is $(a + b)^2/2$ and is taken on when $x = a/\sqrt{2}$, $y = b/\sqrt{2}$.

For our next extension, we go up one dimension and consider the maximum volume parallelepiped whose pair of opposite faces are inscribed in two concentric spheres of radius a and b. Here we want maximize

$$V = 4xy(\sqrt{a^2 - x^2 - y^2} + \sqrt{b^2 - x^2 - y^2}).$$

By the previous rectangle case, $V \leq 8x \sqrt{a^2 - x^2} \cdot \sqrt{b^2 - x^2}$ and by squaring and substitution, this is equivalent to maximizing $t(\alpha - t)(\beta - t)$. Geometrically the latter corresponds to determining the maximum volume of an open box formed from a rectangle 2α by 2β by cutting out squares t by t from the four corners and folding up to form the open box. Some students knowing the A.M.-G.M. inequality would write

$$(\alpha + \beta)/3 = \{2t + (\alpha - t) + (\beta - t)\}/3 \geq \sqrt[3]{2t(\alpha - t)(\beta - t)}$$

which is correct. Unfortunately, to get equality, we must have

$$2t = (\alpha - t) = (\beta - t)$$

and this is only possible if $\alpha = \beta$. Fortunately, an elementary solution is given in a joint paper with R. Boas [27]. We use the weighted A.M.-G.M. inequality with weights to be determined. Here,

$$\{t + r(\alpha - t) + s(\beta - t)\}/3 \geq \sqrt[3]{rst(\alpha - t)(\beta - t)}\,.$$

First the left hand side must reduce to a constant. This requires that $r + s = 1$. Then for equality, we must have $t = r(\alpha - t) = s(\beta - t)$. Eliminating r and s, gives

$$1/(\alpha - t) + 1/(\beta - t) = 1/t \tag{5}$$

which corresponds to the logarithmic derivative of $t(\alpha - t)(\beta - t)$. Another solution can be gotten by using $x^2 + y^2 \geq 2xy$, to give

$$\max V = \max 2\lambda^2 \{\sqrt{a^2 - \lambda^2} + \sqrt{b^2 - \lambda^2}\}.$$

For further extensions, we can obtain the maxima of

$$xyz(\sqrt{a^2 - x^2 - y^2 - z^2} + \sqrt{b^2 - x^2 - y^2 - z^2}),$$
$$xyzw(\sqrt{a^2 - x^2 - y^2 - z^2 - w^2} + \sqrt{b^2 - x^2 - y^2 - z^2 - w^2}),\ldots,$$

in a similar manner or by first using the A.M.-G.M. inequality and in each case the critical value will be determined by a quadratic equation analogous to (5). Again, I leave these as exercises.

VI. A Probability Problem.

In 1950, J.B. Kelly [28] proposed a very interesting dice problem which keeps on resurfacing in different problem journals. He had asked it if were possible to weight a pair of dice so that the probability of each of the numbers $2, 3, \ldots, 12$ coming up were the same. There were two published solutions. The first proof for the weighting being impossible by L. Moser and J. Wahab was sweet and simple. Let a_j and b_j denote the probability of getting a j turning up on each die, respectively, and let P_j denote the probability of getting a sum j turning up for the pair of dice. Then,

$$P_2 = a_1 b_1 = a_6 b_6 = P_{12}.$$

Hence, $(a_1 - a_6)(b_1 - b_6) \leq 0$ and so

$$P_2 + P_{12} = a_1 b_1 + a_6 b_6 \leq a_1 b_6 + a_6 b_1 \leq P_7.$$

The second proof by Finch and Halmos was considerably more sophisticated and uses generating functions (Pattern #9) which originated with Euler. This pattern is generally very useful in determining sequences with prescribed properties. Let

$$F(x) = a_1 x + a_2 x^2 + a_3 x^3 + a_4 x^5 + a_x x^5 + a_6 x^6,$$
$$G(x) = b_1 x + b_2 x^2 + b_3 x^3 + b_4 x^5 + b_5 x^4 + b_6 x^6.$$

It is now natural to form the product

$$F(x) \cdot G(x) = a_1 b_1 x^2 + (a_1 b_2 + a_2 b_1)x^3 + (a_1 b_3 + a_2 b_2 + a_3 b_1)x^4 + \ldots + a_6 b_6 x^{12}$$
$$= (x^2 + x^3 + \ldots + x^{12})/11.$$

Factoring out the x^2, we get

$$(a_1 + a_2 x + \ldots + a_6 x^5)(b_1 + b_2 x + \ldots + b_6 x^5) = (x^{11} - 1)/11(x - 1).$$

The latter equation is impossible for real a_i, b_i since the right hand side only vanishes for complex roots (of unity) while the left-hand side must vanish for at least two real roots.

Having more than one solution for a problem is very desirable since we increase the chances for nontrivial extensions or generalizations. Since one cannot get all the eleven sums with equal probability, it is natural to ask if one can get at least ten of them. Here the first solution does not help but the second solution does. Before we had that

$$(a_1 x + a_2 x^2 + \ldots + a_6 x^6)(b_1 x + b_2 x^2 + \ldots + b_6 x^6) = P_2 x^2 + P_3 x^3 + \ldots + P_{12} x_{12}.$$

If we now let $a_1 = 0$ so that also $P_2 = 0$, we get

$$(a_2 + a_3 x + \ldots + a_6 x^4)(b_1 + b_2 x + \ldots + b_6 x^5) = (1 + x + \ldots + x^9)/10$$
$$= (x^{10} - 1)/10(x - 1)$$
$$= (x^5 - 1)(x^5 + 1)/10(x - 1)$$
$$= (x^5 + 1)/2 \cdot (x^4 + x^3 + x^2 + x + 1)/5.$$

Hence, we have that

$$a_1 = 0, \quad a_2 = a_3 = a_4 = a_5 = a_6 = 1/5,$$

$$b_1 = b_6 = 1/2, \quad b_2 = b_3 = b_4 = b_5 = 0$$

are a possible set of weights. These can be physically achieved by using a pair of regular icosahedra (each of which has 20 equilateral triangle faces). On one of the icosahedra, we number ten faces each with ones and sixes while on the other one we number four faces with each of the numbers 2, 3, 4, 5, 6 as indicated in Figure 12.

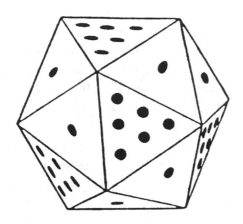 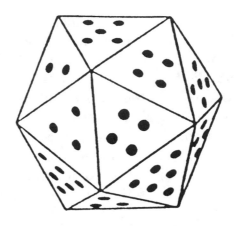

Figure 12

The game known as "craps" is played with two dice (uniform cubes, each with their faces numbered 1 through 6), and the shooter wins unconditionally if he throws a 7 or an 11 (the sum of the top two faces coming up). These points are called "naturals." He loses the game in case he throws a 2,3, or 12 (called "craps"). But if his first throw is a 4,5,6,8,9, or 10, he then throws the dice repeatedly until he throws the same number he had before or until he throws a 7. If he rolls a 7 before obtaining his number, he loses the game; otherwise he wins. It is known that the probability of a shooter winning his throw is slightly less than a half and is given by the following sum:

Point   4,10   5,9   6,8   7   11

$$2\{(3/36) \cdot (3)/(3+6) \ + (4/36) \cdot (4)/(4+6) \ + (5/36) \cdot (5)/(5+6)\} \ + 6/36 \ + 2/36$$

$$= 244/495$$

(Note that the probability a shooter initially throws a 4 is 3/36 and then the probability of throwing another 4 before throwing a seven is 3/(3+6). Hence the probability of initially throwing a 4 and winning is $(3/36) \cdot (3/(3+6))$. The same is true for the point 10. The other probabilities are listed under the respective points.) Coincidentally, the probability of winning in craps with the icosahedral dice is

Point   4,5,6,8,9,10   7   11

$$6\{(1/10) \cdot (1/2)\} + 1/10 + 1/10 = 1/2 \ .$$

Even though icosahedral dice will roll better than cubic dice, these will never be accepted in Las Vegas because the game would now be fair!

VII. The "Navigation" Problem of Zermelo

I am now going to consider an applied problem which is a special case of the navigation problem of Zermelo (of set theory fame). In flying between two cities, airlines want to minimize the time of flight since costs are essentially proportional to the time. Idealizing the problem, the aircraft is to fly in a horizontal plane at a fixed cruising speed (relative to the wind). If there was no wind, the path would be a straight line. Zermelo's problem was, given the wind field as a function of time and space, determine the path of least time. Subsequently, McShane (of integration theory fame) proved the existence of a solution if the wind field was sufficiently smooth. Clearly this is no help to the airlines. They know a solution exists physically, what they need is an algorithm to determine the flight path. Apparently, they do have a computer program which does this and even takes into consideration passenger comfort by avoiding flying through regions of high turbulence. I once tried to get a copy of this program from an airline but I didn't succeed.

I now consider the special case of the problem when the wind field is constant. A pilot friend of mine had given me this problem. At that time he was taking a course on the calculus of variations. He set up the Euler-Lagrange equation and the solution found (just the necessary condition) was a straight line path. Now it is obvious that such a path minimizes the distance flown but it is not immediately obvious that this minimizes the time as well. However, when the end result of a solution which is not particularly simple, is something relatively simple, one should be looking for a simpler solution.

Let me digress to illustrate this point. If, say in a probability problem, the end result is that the probability is $1/2$, then there is likely to be symmetry argument for it. As a specific example [29-30], suppose Mary and John toss $n+1$ and n fair coins, respectively, what is the probability that Mary gets more heads than John. One expects the answer to be a function of n and one can obtain the answer by setting up a double sum of products of binomial coefficients in a straight forward manner. But by considering the special cases $n = 0, 1, 2$, which are quite easy to calculate, the probability is $1/2$. I leave it as an exercise to find the easy symmetry solution for general n. This will then give the value of the above double sum (Pattern #16).

Returning to our navigation problem, we can simplify the problem by choosing an appropriate coordinate system. We take our origin in the moving wind field. This is equivalent to imparting the negative wind velocity into the system of plane and destination city. We now have equivalently that the airplane is flying at its constant cruising speed and the destination is moving with the negative wind velocity. Now assume any path of the airplane that intersects the moving city in both time and space as in Figure 13.

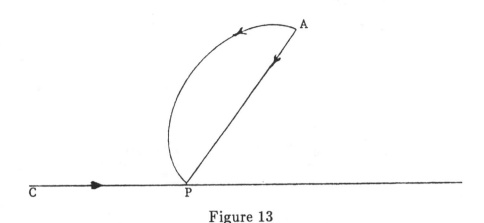

Figure 13

Now consider any straight path of the plane to this spatial intersection point P. Since the latter path is a straight line it is shorter in length than the curved path and consequently the time of arrival at P would be earlier. For other applications of moving coordinate systems, see [31-32].

VIII. Miscellaneous Transformations

To conclude this paper, I will expand upon Pattern #18h, Miscellaneous Transformations. In fact all of #18 and some earlier ones e.g., generating functions, equivalence classes, could be lumped under "transforms" in general and this encompasses a large part of mathematics. Since this has been treated before with many examples in a joint paper [33] with D.J. Newman, I will only give a quick sketch here. Imagine a mathematician faced with a difficult problem which he either cannot solve or else can only solve with difficult horrendous calculations (to be sure some of these can now be simplified with computer programs like Mathematica and Maple). If he is a hard worker he may succeed in reducing the problem to an easier one. This reduction has changed or transformed the hardness out of the original problem and hence the name "transform." We might picture the mathematician as wishing to go from the box on the left to the box on the right in Figure 14, and finding himself unable or unwilling to do so directly, he takes the circuitous path in Figure 15. The three arrows in Figure 15 might be called in turn.

    I. TRANSFORM,

    II. SOLVE,

    III. INVERT.

Figure 14

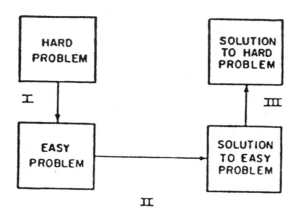

Figure 15

In ancient times, the multiplication of two numbers was something practiced only by the geniuses of the day. No wonder either, the multiplication being carried out in Roman numerals. Today a grade-school boy can (presumably) easily knock off

$$(LXXIV) \quad times \quad (XXVIII)$$

for this is simply

    I. $74 \times 28;$

then

    II.

$$\begin{array}{r} 74 \\ \times \quad 28 \\ \hline 592 \\ 148 \\ \hline 2072 \end{array}$$

and finally,

    III. $2072 = MMLXXII.$

This schoolboy (who probably can multiply better than we do) has already grasped the whole philosophy of transform theory. He took the three steps:

I. TRANSFORM,   II. SOLVE,   III. INVERT,

in a very clearly delineated fashion.

Even this simple little example brings out one of the basic properties of a transform. A transform corresponds to a representation. The transform in this case amounted to representing the Roman numbers in the ordinary Hindu-Arabic system. On a much higher level, e.g., the Fourier transform corresponds to the representation

$$F(x) = \frac{1}{2\pi} \int_{-\infty}^{\infty} e^{-itx} \int_{-\infty}^{\infty} e^{ist} F(s)\,ds\,dt$$

and so on for other familiar transforms. Thus a definition of transform theory might be: The use of a new representation to make a hard problem simpler. Perhaps in the future, one may find a suitable representation to simplify the solution of the four color problem.

References

[1] Golomb, S.W., *Polyominoes*, Charles Scribner's Sons, New York, 1965.

[2] Eves, H., "Philo's Line," Scripta Math. **24** (1959), 141-148.

[3] Klamkin, M.S., "The Olympiad Corner: 80," Crux Mathematicorum **12** (1986), 263-281.

[4] Pólya, G., *How to Solve It*, Princeton University Press, Princeton, 1973.

[5] Pólya, G., *Mathematical Discovery*, Wiley, New York, 1981.

[6] Pólya, G., *Mathematics and Plausible Reasoning*, I, II, Princeton University Press, Princeton, 1954.

[7] Hilbert, D., "Mathematical Problems," Bull. Amer. Math. Soc. **8** (1901-02) 437-479.

[8] Campbell, D. M., and D. M. Higgins, *Mathematics-People-Problems-Results*, I, Wadsworth, Belmont, 1984, 273-278, 300-304.

[9] Browder, F.E., ed., *Mathematical Developments Arising from Hilbert Problems*, I, II, Amer. Math. Soc., Providence, 1976.

[10] Klamkin, M.S., ed., *Problems in Applied Mathematics*, SIAM, Philadelphia, 1990.

[11] Hadamard, J., *The Psychology of Invention in the Mathematical Field*, Dover, New York, 1945.

[12] Klamkin, M.S., "On the Teaching of Mathematics so as to be Useful," Educ. Studies Math. **1** (1968), 126-160.

[13] Niven, I., *Maxima and Minima Without Calculus*, M.A.A., Washington, D.C., 1981.

[14] Klamkin, M.S., "On Two Classes of Extremal Problems Without Calculus," Math. Mag. (in press).

[15] Kazarinoff, N.D., *Geometric Inequalities*, Random House, New York, 1961.

[16] Albers, D.J., "Interview with Irving Kaplansky," College Math. Jour. **22** (1991), 99-117.

[17] Klamkin, M.S., "Vector Proofs in Geometry," Amer. Math. Monthly **77** (1970), 1051-1065.

[18] Gleason, A. M., Greenwood, R. E. and L. M. Kelly, *The William Lowell Putnam Competition*, M.A.A., Washington, D.C., 1980.

[19] Klamkin, M.S., USA Mathematical Olympiads 1972-1986, M.A.A., Washington, D.C., 1988.

[20] Wiener, N., "The Shortest Line Dividing an Area in a Given Ratio," Proc. Philo. Soc. **18** (1914), 56-58.

[21] Croft, H.T., Falconer, K.J. and R.K. Guy, *Unsolved Problems in Geometry*, Springer-Verlag, New York, 1991.

[22] Klamkin, M.S., "On Some Symmetric Sums of Unit Vectors," Math. Mag. **64** (1991), 271-273.

[23] Eves, H., "Maxima and Minima by the Method of Coincidence of Equal Values with an Application to Philo's Line," Crux Mathematicorum **6** (1980), 232-236.

[24] Problem 488, Crux Mathematicorum **6** (1980), 260-263.

[25] Klamkin, M.S., and A. Meir, "Extensions of an Elementary Inequality," Aequationes Math. **26** (1983), 197-207.

[26] Crux Mathematicorum **11** (1985), 115, 218.

[27] Boas, R.P., and M.S. Klamkin, "Extrema of Polynomials," Math. Mag. **50** (1977), 75-78.

[28] Kelley, J.B., Problem E925, Amer. Math. Monthly **58** (11951), 191-191.

[29] Klamkin, M.S., "A Probability of More Heads," Math. Mag. **44** (1971), 146-149.

[30] Klamkin, M.S., "Symmetry in Probability Distributions," Math. Mag. **61** (1988), 193-194.

[31] Klamkin, M.S., and D.J. Newman, "Flying in a Wind Field, I, II," Amer. Math. Monthly **76** (1969), 16-23, 1013-1019.

[32] Klamkin, M.S., "Moving Axes and the Principle of the Complementary Function," SIAM Review **16** (1974), 295-302.

[33] Klamkin, M. S., and D. J. Newman, "The Philosopy and Applications of Transform Theory," SIAM Review **3** (1961), 10-36.

Bi-centric Quadrilaterals and the Pedal *n*-gon

Gordon Bennett
Florida Keys Community College
Key West, Florida 33041

Introduction

E.W. Hobson [1] gives the following problem: (see figure 1)

A quadrilateral is such that a circle can be described about it and another inscribed in it; shew the radius of the latter is

(1)
$$\frac{2\sqrt{abcd}}{a+b+c+d}.$$

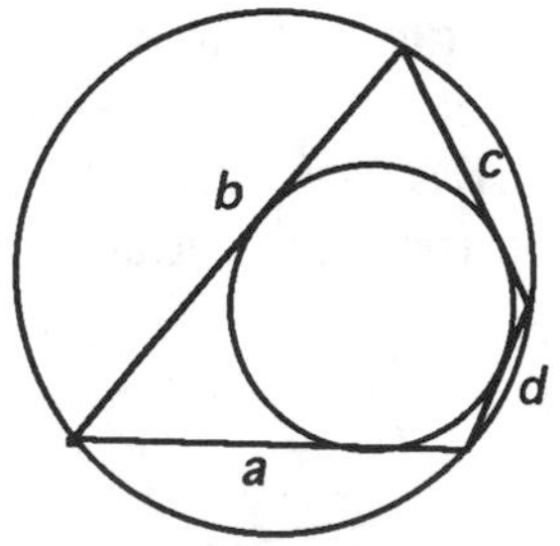

Figure 1

N.A. Court [2] calls the cyclic, inscribed quadrilateral **bi-centric,** and asks the reader in problem 6, page 139, to show the square of its area is the product of the four sides. R. Johnson [3] gives an easy way to construct a bi-centric quadrilateral in the following theorem:

(2)　If a quadrilateral is circumscribed about a circle, its vertices lie on another circle if and only if the lines joining the points of contact of opposite sides are mutually perpendicular.

Byerly [4] gives the following theorem.

A necessary and sufficient condition that, given two circles (radii, r, R), a quadrilateral can be described about one and inscribed in the other is

(3)　　　　　$r^{-2} = (R+d)^{-2} + (R-d)^{-2}$

where d is the distance between the centers of the two given circles.

97

Equation (3) is derived very neatly in [3, p. 95]. I will also contribute a formula for this distance in section 3.

H.S.M. Coxeter and S.L. Gretzer [5] give the following (in a footnote):

> "According to J. L. Coolidge, 'A Treatise on the Circle and the Sphere', (Oxford, 1916), pp. 45-46, it was Euler who discovered this as well as the analogous formula
>
> (4) $$r^{-1} = (R+d)^{-1} + (R-d)^{-1}$$
>
> for a triangle."

They also give a very simple formula for the *inversive* distance between the two centers. While almost all authors ascribe (4) to Euler, it has been pointed out by MacKay that it was discovered by Chapple. (J.S. MacKay, "Historical Notes on a Geometric Theorem and Its Development," Proc. Edinburgh Math Soc. 5(1887) 6263.)

More recently, I found the following problem in the 41st Annual High School Mathematics Examination.

> A quadrilateral that has consecutive sides of length 70, 90, 130 and 110 is inscribed in a circle and also has a circle inscribed in it. The point of tangency of the inscribed circle to the side of length 130 divides that side into segments of length x and y. Find $|x-y|$.
>
> (A) 12 (B) 13 (C) 14 (D) 15 (E) 16

My approach is to start with a triangle ΔA_i and its sides a_i $(i = 1,2,3)$, along with its circumcircle with center O, and radius R. Let P be some point on a_3. Then, A_3P (extended) will intersect the circumcircle at, say, B_3. Let $A_1B_3 = b_1$ and $A_2B_3 = b_2$. If B_3 is located just right, a circle with center o' and radius r' may be inscribed, touching the sides of the quadrilateral at N_i $(i = 1,2,3,4)$. (See Figure 2).

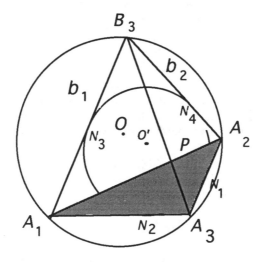

Figure 2

A theorem, giving the sides b_1 and b_2 in terms of ΔA_i, will be proved and this will lead to another construction for a bi-centric quadrilateral. Areal coordinate techniques will be used to develop some data which will then be used to prove several theorems relative to quadrilateral N_i.

The main section on pedal *n*-gons ($n = 3, 4$) will then be introduced along with some proofs and an interesting conjecture. The final section gives an example that combines both of the main topics mentioned in the title. I will conclude with a few comments and the answer to the above problem.

1. Theorem to determine b_1 and b_2

The sides b_1, b_2 are given in terms of the sides a_i of ΔA_i as follows.

$$(5) \qquad b_1 = \frac{a^2 - a_1 + \sqrt{W}}{2}$$

$$(6) \qquad b_2 = \frac{a_1 - a_2 + \sqrt{W}}{2}$$

where s = semi-perimeter of ΔA_i and

$$W = \frac{a_1 a_2 a_3^2 - (s - a_1)(s - a_2)(a_1 - a_2)^2}{s(s - a_3)}$$

Proof:

Starting with $b_1 - b_2 = a_2 - a_1$ and $a_3^2 = b_1^2 + b_2^2 + 2b_1 b_2 \cos A_3$, squaring the first equation and subtracting from the second gives:

$$4b_1 b_2 = \frac{a_3^2 - (a_2 - a_1)^2}{\cos^2 1/2 A_3}$$

and adding this to $(b_1 - b_2)^2 = (a_2 - a_1)^2$ gives

$$(b_1 + b_2)^2 = \frac{a_3^2 - (a_1 - a_2)^2 \sin^2 1/2 A_3}{\cos^2 1/2 A_3}.$$

Equations (5) and (6) follow immediately.

2. A Euclidean construction for the bi-centric quadrilateral

Start with ΔA_i and its circumcircle. For ease in describing the construction, label ΔA_i so $a_2 > a_1$.

1)   Find E on the angle bisector of A_3 so that $A_3 E = a_2 - a_1$.
2)   Drop a perpendicular from E to F on side $A_1 A_3$.
3)   On a semi-circle constructed on side $A_1 A_2$, find a point G such that $A_1 G = EF$.
4)   On side $A_1 A_3$ (extended if necessary) find a point H such that $A_3 H = A_2 G$.
5)   Let the perpendicular at H intersect the angle bisector $A_3 E$ (extended) at K. The side b_2 of our bi-centric quadrilateral $= \frac{1}{2} EK$.

6) B_3 is now **determined**. Let $A_3 B_3$ intersect $A_1 A_2$ at P. Draw PO intersecting the angle bisector $A_3 E$ at o'. An alternate way to determine o' that does not use the center O is to join the midpoints of the interior diagonals of the quadrilateral $A_1 B_3 A_2 A_3$. This line will intersect the angle bisector at o' also.

7)   The inscribed circle may now be drawn.

3.   General distance and angle data for the bi-centric quadrilateral

Let $A_1 A_2 A_3$ be the reference triangle. Then the respective **areal** coordinates of P and B_3 are

$$(a_1 b_2, a_2 b_1, 0)/k_1 \text{ and } (a_1 b_2 k_3, a_2 b_1 k_3, -b_1 b_2 k_1)/a_1 a_2 k_1 \text{ where}$$
$$k_1 = a_1 b_2 + a_2 b_1, \ k_2 = a_1 b_1 + a_2 b_2, \text{ and } k_3 = a_1 a_2 + b_1 b_2.$$

Now assume that B_3 is such that $A_1 B_3 A_2 A_3$ is bi-centric. We can determine the areal coordinates of the N_i as a function of the ratios that each N_i divides its corresponding side. It can be shown that $N_1 A_3 / A_2 N_1 = a_2 / b_2$ so that the areal coordinates of N_i are $(0, a_2, b_2)/S$ where $S = a_1 + b_1 = a_2 + b_2$. In a similar fashion we can obtain the respective areal coordinates of N_2, N_3, N_4, as

$$(a_1, 0, b_1)/S$$
$$(a_1 b_2 S^2, a_2 b_1 k_3, -b_1 b_2 k_1)/a_1 k_1 S,$$
$$(a_1 b_2 k_3, a_2 b_1 S^2, -b_1 b_2 k_1)/a_2 k_1 S.$$

The sines and cosines of several angles for the quadrilateral as well as various distances may now be given in terms of the following.

A' = area $A_1 B_3 A_2 A_3 = \sqrt{a_1 a_2 b_1 b_2}$, and $r' = A'/S$. (see **Introduction** (1))

β = one of the angles formed by the diagonals $A_3 B_3$ and $A_1 A_2$,

$\varnothing$ = say, $\angle B_3 A_2 A_3$

Also, let

$$k_1' = a_1 b_2 - a_2 b_1, \ k_2' = a_1 b_1 - a_2 b_2, \ k_3' = a_1 a_2 - b_1 b_2 \text{ and } k = \frac{4A'^2}{k_1 k_3}.$$

Then

$$\sin \varnothing = \frac{2A'}{k_1}, \qquad\qquad \cos \varnothing = \frac{k_1'}{k_1}.$$

$$\sin \beta = \frac{2A'}{k_2}, \qquad\qquad \cos \beta = \frac{k_2'}{k_2}.$$

$$\sin A_3 = \frac{2A'}{k_3}, \qquad\qquad \cos A_3 = \frac{k_3'}{k_3}.$$

Depending on the angle chosen and the shape of the quadrilateral, the above data gives the angle or its supplement. The following distances may also be derived.

$$A_1P = \frac{a_2b_1\sqrt{1+k}}{S}, \qquad N_1P = r'\sqrt{\frac{a_1k}{b_1}}, \qquad N_1N_2 = \frac{2a_1a_2}{S}\sqrt{\frac{b_1b_2}{k_3}}.$$

$$A_2P = \frac{a_1b_2\sqrt{1+k}}{S}, \qquad N_2P = r'\sqrt{\frac{a_2k}{b_2}}, \qquad N_2N_3 = \frac{2a_2b_1}{S}\sqrt{\frac{a_1b_2}{k_1}}.$$

$$A_3P = \frac{a_1a_2\sqrt{1+k}}{S}, \qquad N_3P = r'\sqrt{\frac{b_1k}{a_1}}, \qquad N_3N_4 = 2\frac{b_1b_2}{S}\sqrt{\frac{a_1a_2}{k_3}}.$$

$$B_3P = \frac{b_1b_2\sqrt{1+k}}{S}, \qquad N_4P = r'\sqrt{\frac{b_2k}{a_2}}, \qquad N_4N_1 = 2\frac{a_1b_2}{S}\sqrt{\frac{a_2b_1}{k_1}}.$$

Finally, $N_1N_3 = \sqrt{a_2b_2k}$, $N_2N_4 = \sqrt{a_1b_1k}$ and

$$d = \boxed{R\sqrt{\frac{1-k}{1+k}}} \quad \text{(see \textbf{Introduction} (3))}$$

$$= \text{distance between the centers of the bi-centric circles.}$$

4.   **Three theorems for quadrilateral N_i** (see Figure 3)

A)   The diagonals of quadrilateral N_i *intersect* at P and are *perpendicular* for (see **Introduction** (2))

$$PN_1 + PN_3 = N_1N_3, \text{ etc. and } PN_1^2 = N_1N_2^2, \text{etc.}$$

B)   The (interior) diagonals of quadrilateral N_i *bisect* the angles formed by the (interior) diagonals of quadrilateral $A_1B_3A_2A_3$ for

$$A_1N_2/N_2A_3 = A_1P/PA_3, \text{ etc.}$$

C)   Quadrilateral N_i *loses* the property of "being inscribed" (unless $k = 1$) for

$$N_1N_2 + N_3N_4 = 2r'\frac{\sqrt{a_1a_2} + \sqrt{b_1b_2}}{\sqrt{k_3}} \text{ and}$$

$$N_2N_3 + N_4N_1 = 2r'\frac{\sqrt{a_1a_2} + \sqrt{b_1b_2}}{\sqrt{k_1}}.$$

Hence, quadrilateral N_i, to be inscribed, requires that $k_1 = k_3$ which implies that $(a_1 - b_1)(a_2 - b_2) = 0$. Thus, since quadrilateral $A_1B_3A_2A_3$ is cyclic, $a_1 = b_1$ implies a_2 must equal b_2 and the only rectangles that can be inscribed are squares.

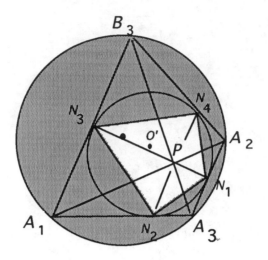

Figure 3

5. Pedal *n*-gons and a conjecture

It is known [6,7] that, for a given point in the *n*-gon plane, the *n*th pedal *n*-gon of any *n*-gon is similar to the original *n*-gon. If $n = 3$, [5 pp. 22-25] gives a good discussion on pedal triangles and the "parade" of angles involved.

Let P be any point in the interior of the triangle ΔA_i ($i = 1,2,3$), with sides a_i, circumradius R, center O and area F. Let $R_i = PA_i$ and r_i denote the perpendicular distance from P to the sides a_i. Let $d_p^2 = R^2 - PO^2$ and F_n denote the area of the *n*th pedal *triangle*. Then, (7), (8) and (9) are, respectively

$$(7)\qquad F_1/F = d_p^2 \big/ 4R^2$$

$$(8)\qquad F_2/F = r_1 r_2 r_3 d_p^4 \big/ 2R\left(R_1 R_2 R_3\right)^2$$

$$(9)\qquad F_3/F = \left(r_1 r_2 r_3 / R_1 R_2 R_3\right)^2.$$

A derivation of (7) is given in [1, pp. 209-210] and [3,p.139]. Apparently, (8) and (9) are new. As special cases, when P is the incenter of $A_1 A_2 A_3$,

$$F_1/F = r/2R, \quad F_2/F = r/8R \qquad\qquad F_3/F = \left(r/4R\right)^2$$

where r is the inradius. When P is the circumcenter O,

$$F_1/F = 1/4 \qquad F_2/F = \cos A_1 \cos A_2 \cos A_3 /2. \qquad F_3/F = \left(\cos A_1 \cos A_2 \cos A_3\right)^2.$$

Note: $\dfrac{r_1 r_2 r_3}{R_1 R_2 R_3}$ is invariant under an isogonal transformation. Therefore

$$F_3/F = \left(\cos A_1 \cos A_2 \cos A_3\right)^2 \text{ when } P \text{ is the orthocenter } H \text{ also.}$$

As a possible extension of (9), I **conjecture** that:

(10)
$$\boxed{\frac{F_n}{F} = \left(\frac{r_1 r_2 r_3 \cdots r_n}{R_1 R_2 R_3 \cdots R_n}\right)^2}$$

where F_n and F are *now* the areas of the nth pedal n-gon and the initial n-gon respectively, and r_i and R_i are the distances from P to sides and vertices, respectively, of the initial n-gon. Equation (10) may be cast in the alternate form (see figure 4 for an example when $n = 4$)

(11)
$$\boxed{\frac{F_n}{F} = \prod_{i=1}^{n} \sin^2 \varnothing_i}.$$

where $\varnothing_i = \angle PA_i A_{i+1}$ and $A_{n+1} \equiv A_1$.

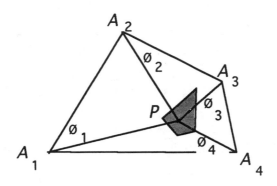

Figure 4

To prove (7), (8) and (9), let the four triangles be denoted by ΔA_i, ΔB_i, ΔC_i and ΔD_i with corresponding sides a_i, b_i, c_i and d_i (see figure 5). Let $\angle A_2 P A_3 = A_i'$, etc. Then the angles of the four triangles are A_i, $A_i' - A_i$, $\pi - A_i'$, A_i, respectively. Clearly $\sin A_1' = a_1 r_1 / R_2 R_3$, etc. An inversion transformation on the sine law $a_i / \sin A_i = 2R$ yields

$$\frac{a_i R_i}{\sin A_i' - A_i} = \frac{2 R R_1 R_2 R_3}{d_p^2}.$$

Thus, $\sin(A_1' - A_1) = a_1 d_p^2 / 2 R R_2 R_3$, etc. Hence $b_1 = R_1 \sin A_1 = R_1 a_1 / 2R$, etc., and

$$\text{Area } \Delta B_i = 1/2 b_2 b_3 \sin(A_1' - A_1) = d_p^2 F / 4 R^2$$

where F denotes the area of the original triangle ΔA.
Similarly,

$$c_1 = R_1 (\text{relative to } \Delta B) \sin(A_1' - A_1) = r_1 (\text{relative to } \Delta A) \sin(A_1' - A_1) = a_1 r_1 d_p^2 / 2 R R_2 R_3$$

and hence

$$\text{Area } \Delta C_i = \frac{1}{2} c_2 c_3 \sin(\pi - A_1') = r_1 r_2 r_3 d_p^4 F / 2 R R_1^2 R_2^2 R_3^2.$$

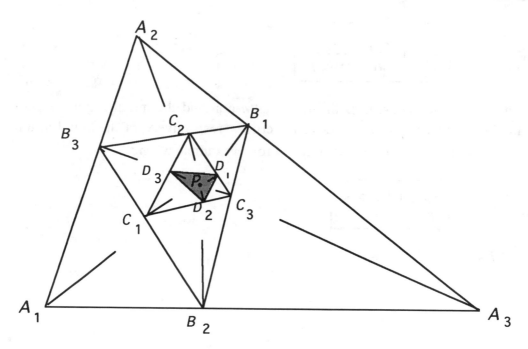

Figure 5

Finally,

$$d_1 = R_1(\text{relative to } \Delta C)\sin\left(\pi - A_1'\right)$$
$$= r_1(\text{relative to } \Delta B)\sin A_1'$$

but

$$b_1 r_1(\text{relative to } \Delta B) = r_2 r_3 \left(\sin\left(\pi - A_1\right)\right)(\text{relative to } \Delta A).$$

Therefore,

$$r_1(\text{relative to } \Delta B) = r_2 r_3 / R_1 (\text{relative to } \Delta A)$$

Hence $d_1 = r_1 r_2 r_3 \, a_1 / R_1 R_2 R_3$ and Area $\Delta D = d_1^2 F / a_1^2$ follows immediately.

6. Final section:

Consider now the first four pedal quadrilaterals generated by the point o' on the bi-centric quadrilateral $A_1 B_3 A_2 A_3$. The first pedal quadrilateral is just N_i. If, again from o', the second quadrilateral, say M_i, is formed, the M_i's are just the midpoints of the quadrilateral N_i and form a rectangle (see figure 6). Continuing in this same way, we get the third and fourth pedal quadrilateral, say Q_i and R_i. Then the following area relationships may be derived. (Here $A' = F$, F_1 = area of quadrilaterial N_i, F_2 = area of quadrilaterial M_i, etc.)

$$F_1/F = \frac{1}{2}k, \quad F_2 / F = \frac{1}{4}k, \quad F_3 / F = \frac{1}{8}k, \quad F_4 / F = \left(\frac{1}{4}k\right)^2$$

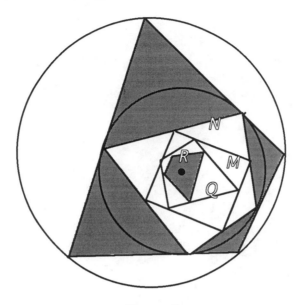

Figure 6

Clearly, this nested "parade" continues, with every fourth quadrilateral being bicentric.

7. Summary:

Recently, I transcribed parts of this article into the mathematical computer program *Theorist*. Areal coordinates were found useful and easily translate to the rectangular coordinate system. *Theorists* ability to animate geometric figures gives some of the geometry in this paper an amazing *dynamics*.

The answer to the 41st AHSME problem is, of course, **(B)**.

[This author wishes to thank the referees for many helpful suggestions that improved this paper].

References

1. E.W. Hobson, *A Treatise on Plane & Advanced Trigonometry*, Dover (7th ed.), 1928, New York, pp. 206-9.

2. N.A. Court, *College Geometry*, Barnes & Nobel (2nd ed.), 1952, New York, p. 139.

3. R.A. Johnson, *Advanced Euclidean Geometry*, Dover, 1960, New York, p. 95, pp. 139-40.

4. W.E. Byerly, The In-and-Circumscribed Quadrilateral, *Annals of Mathematics*, 10 (1909), 123-8.

5. H.S.M. Coxeter, S.L. Greitzer, *Geometry Revisited*, Random House, 1967, New York, pp. 22-25, p.130.

6. A. Oppenheim, The Erdös Inequality and Other Inequalities for a Triangle, *American Mathematical Monthly*, 68 (1961), 230.

7. B.M. Stewart, Cyclic Properties of Miquel Polygons, *American Mathematical Monthly*, 47 (1940), 462.

The Fundamental Group as a Beautiful Functor

Peter Hilton
SUNY Binghamton
Binghamton, New York 13902-6000

Dedicated to Howard Eves on the occasion of his 80th birthday

1. INTRODUCTION

The aspect of Howard Eves' many-faceted contributions to mathematics which I propose to highlight in this talk is his strong sense of the esthetic quality inherent in all good mathematics and in the effective presentation, oral or written, of mathematics. This esthetic quality is frequently a guiding, stimulating force in the origination and development of new mathematical ideas; the universality of this proposition is attested by the testimony of the applied mathematician Roger Penrose [1], who regards esthetics as the principal motive force in mathematical research, even where a gain in scientific knowledge is an anticipated outcome of that research.

In presenting mathematics we must provide convincing evidence of both its power and its beauty. The study of geometry is excellently adapted to the demonstration of beauty in mathematics; but power is more obviously associated with the systematic parts of mathematics – for example, arithmetic, algebra, analysis – than with its more inspirational aspects.[1] Thus I have been led to conclude that we demonstrate the true, expanded nature of mathematics when we show how algebraic methods may be applied to the study of geometry.

Such passages from geometry to algebra – or, more generally, from one part of mathematics to another – are called *functors* [2]; and I wish to talk about just one such functor today, namely, the *fundamental group*. This was invented by the great French mathematician Henri Poincaré in the course of his studies of differential equations, and is known to play a crucial role in various parts of mathematics, for example, complex variable theory. However, I shall only consider here its role as a link between topology and group theory.

---

[1] To prove theorems in pure synthetic geometry, one needs a new idea for each problem. Such a high idea-to-problem ratio can only be sustained by the consummate artist; and no human activity can become popular if one must be a skilled and talented artist to engage in it.

We will consider a restricted category of topological spaces, namely, that of connected polyhedra; moreover we will assume our spaces *pointed* in the sense that each is furnished with a base point, which may be taken to be a vertex of the underlying combinatorial structure of the polyhedron. Thus we write τ for our topological category and simply call an object X of τ a *space*. The fundamental group functor attaches to each space X a group πX and to each (based) continuous function $f: X \longrightarrow Y$ a group homomorphism $\pi f: \pi X \longrightarrow \pi Y$, such that the following three conditions are satisfied.

(i) $\pi Id = Id$, where Id is an identity (function or homomorphism)

(ii) $\pi(gf) = (\pi g)(\pi f)$, for $f: X \longrightarrow Y$, $g: Y \longrightarrow Z$

(iii)   if f is continuously deformable to f' with no movement of basepoint, then $\pi f = \pi f'$

In Section 2 we will give the explicit description of π and show in Section 3 how it may be calculated. In so doing we will introduce the notion of the *edge fundamental group* of a simplicial complex, thus forging an important link between group theory and combinatorics. Then in Section 4 we will endeavor to explain our claim that π is a beautiful functor; and give examples to show how certain classical theorems may be proved using the elegant properties of π.

2. DEFINITION OF THE FUNDAMENTAL GROUP

Let X be a pointed space with base point x_0. A *loop* on X is a continuous function ℓ from the unit interval I, given by $0 \le t \le 1$, to X, such that $\ell(0) = \ell(1) = x_0$. Given two loops ℓ, m we may *compose* them to form $n = \ell \circ m$, given by

$$n(t) = \left\{ \begin{array}{ll} \ell(2t), & 0 \le t \le \frac{1}{2} \\ m(2t-1), & \frac{1}{2} \le t \le 1 \end{array} \right.$$

A *homotopy*, or deformation, of the loop ℓ, is a continuous function $H: I \times I \longrightarrow X$ such that $H(t, 0) = \ell(t)$ and $H(0, u) = H(1, u) = x_0$, $\forall \ u \in I$. If ℓ' is the loop given by $\ell'(t) = H(t, 1)$, we say that ℓ' is *homotopic* to ℓ and write $\ell' \simeq \ell$ or $\ell' \overset{H}{\simeq} \ell$. It is then easy to show that $\simeq$ is an equivalence relation, and that

(a)   if $\ell' \simeq \ell$, $m' \simeq m$, then $\ell' \circ m' \simeq \ell \circ m$;

(b) $\ell \circ c \simeq \ell \simeq c \circ \ell$, where c is the constant loop given by $c(t) = x_0$;

(c) $\ell \circ \bar{\ell} \simeq c \simeq \bar{\ell} \circ \ell$, where $\bar{\ell}$ is the loop obtained by reversing ℓ, that is,
$\bar{\ell}(t) = \ell(1-t)$;

(d) $(\ell \circ m) \circ n \simeq \ell \circ (m \circ n)$.

These facts show that the set of homotopy classes of loops on X constitute a group under the binary operation induced by composition of loops; that is the fundamental group, which we write πX. Note that, as defined, πX depends on the choice of base point in X; however, one shows that different choices of base point lead to isomorphic groups.

Now let $f\colon X \longrightarrow Y$ be a (based) continuous function. Then, if ℓ is a loop on X, $f\ell$ is a loop on Y. Moreover,

(e) $f(\ell \circ m) = f\ell \circ fm$;

(f)   if $\ell \simeq \ell'$ then $f\ell \simeq f\ell'$;

(g)   if there is a based homotopy from the function f to the function f', then $f'\ell \simeq f\ell$;

(h) $(gf)\ell = g(f\ell)$.

Properties (e) and (f) show that f induces a homomorphism $\pi f\colon \pi X \longrightarrow \pi Y$. Property (i) of πf is obvious; property (ii) follows from property (h); and property (iii) from property (g).

The fundamental group may loosely be described as counting the (2-dimensional) holes in the space X. Thus if X is a circle, πX is cyclic infinite; if X is a torus, πX is free abelian on 2 generators; if X is a figure **8**, πX is free on 2 generators; if X is a polyhedron in the geometer's usual, restricted sense (a homeomorph of the 2-dimensional sphere), πX is trivial since X has a 3-dimensional hole, but not a 2-dimensional hole. However, πX is really much more subtle than this — but to show its subtlety we must introduce a way of calculating it from the combinatorial structure of X.

3. CALCULATING THE FUNDAMENTAL GROUP

We are assuming that our space X is a polyhedron; more specifically, we will suppose that X is the underlying space of a simplicial complex K and that, moreover, the base point x_0 is a vertex of K. We will associate with the pointed complex K a group $\pi_1 K$ and describe how to set up an isomorphism between $\pi_1 K$ and πX; we call $\pi_1 K$ the *edge fundamental group of K*.

In fact, the development of our definition of $\pi_1 K$ parallels that of the definition of πX. Thus a *loop* on K is a sequence $x_0 x_1 \cdots x_n x_0$ of vertices of K such that, for all i, $0 \le i \le n$, x_i and x_{i+1} are vertices of a simplex[2] of K ($x_{n+1} = x_0$). We compose two loops

$$\ell = x_0 x_1 \cdots x_n x_0 \quad \text{and} \quad \ell' = x_0 y_1 \cdots y_m x_0$$

to form the loop $\qquad \ell * \ell' = x_0 x_1 \cdots x_n x_0 y_1 \cdots y_m x_0$.

We introduce a notion of *edge homotopy* as follows. Let x, x', x'' be vertices of a simplex[3] of K; then we may replace xx'' by $xx'x''$, or make the reverse replacement, in the expression for a loop. Such a replacement is called an *allowed move*; and an edge homotopy is a (finite)

---

[2] Thus they are vertices of an edge of K if they are distinct.

[3] Thus they are vertices of a triangle of K if they are distinct.

sequence of allowed moves. Propositions corresponding to those described in connection with πX now show that the edge homotopy classes of loops on K form a group under the operation induced by composition of loops. However, much more is true. There is an evident *realization functor* from simplicial complexes to polyhedra and then the following diagram commutes

$$(3.1)$$

This means that we have only to calculate $\pi_1 K$. A nice technique for doing this is the following. We may always choose, for any given (connected) K, a subcomplex L such that (i) L is 1-connected, meaning that $\pi_1 L$ is a trivial group, and (ii) L contains all the vertices of K. For example, we may take L to be a maximal tree in K. Then we define a group $G(K, L)$ as follows.

The generators g^{ij} of $G(K, L)$ are in one-one correspondence with the edges $a_i a_j$ of K. There is a relation $g^{ij} = 1$ if $a_i a_j \in L$.

There is a relation $g^{ij} g^{jk} = g^{ik}$ if a_i, a_j, a_k span a simplex of K.

Then

THEOREM $G(K, L) \cong \pi_1 K$.

EXAMPLES   (i)   The circle is a homeomorph of a hollow triangle. Thus, to compute the fundamental group of a circle we take

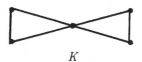

Thus $G(K, L)$ has one generator, corresponding to $a_0 a_1$, and no relations. Thus π(circle) is a cyclic infinite group.

(ii)   The figure **8** may be triangulated as

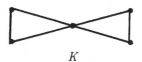

We choose

as L, so that $G(K, L)$ has two generators and no relations. Thus $\pi\left(\text{figure } \mathbf{8}\right)$ is free on 2 generators.

(iii)   A (two-dimensional) sphere may be triangulated as the boundary of a tetrahedron

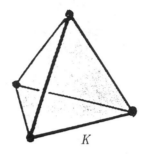

K

We may take as L the whole of K with the interior of one triangle removed; for the resulting space is a cone on a triangular base and so certainly contractible. But every edge of K is in L, so $G(K, L)$ is the trivial group. Thus the sphere is 1-connected.

We close this section with an important remark about our diagram (3.1). Using our Theorem it is not difficult to show that, given any group G, there exists a two-dimensional[4] complex K such that $\pi_1 K = G$. This gives us scope to develop *combinatorial group theory*, which is today a very active subject. For we may mirror the key constructions of group theory (subgroups, normal subgroups, quotient groups, . . .) in the category of simplicial complexes (by means of covering complexes, regular covering complexes, cover transformation groups, . . .). I will indicate just one beautiful application of combinatorial methods. It is easy to see, from our Theorem, that a group G is free if and only if there is a complex K which is 1-dimensional and such that $\pi_1 K = G$. Now subgroups correspond to covering complexes and it is evident that a covering complex $\tilde{K}$ of K has the same dimension as K. Thus we have proved

THEOREM   *A subgroup of a free group is free.*

4. SPECIAL FEATURES OF THE FUNDAMENTAL GROUP FUNCTOR

Among the most important constructions in a mathematical category are the *product* and *coproduct*. The product of two objects A and B is usually the cartesian product $A \times B$ of the underlying sets, together with a structure appropriate to the category. We may formalize the requirements on the product P of A and B as follows. We call the functions appropriate

_____
[4] Or possibly even 1-dimensional — see our first two examples above.

to a particular category the *morphisms* of that category; thus the morphisms of the category of groups are homomorphisms, the morphisms of the category of topological spaces are continuous functions, . . .

We say that the triple $(P; p_A, p_B)$ is the *product* of A and B if

(i) $p_A: P \longrightarrow A$, $p_B: P \longrightarrow B$ are morphisms; and

(ii)   given any object X and morphisms $f_A: X \longrightarrow A$, $f_B: X \longrightarrow B$, there exists a unique morphism $f: X \longrightarrow P$ such that $p_A f = f_A$, $p_B f = f_B$.

The appropriate diagram is

$$(4.1)$$

We will simply write $A \times B$ for P. It is then easy to see that the functor π from pointed spaces to groups is product-preserving. We confine ourselves to polyhedra and assert:

THEOREM P $\pi: \boxed{Polyhedra} \longrightarrow \boxed{Groups}$ *is product-preserving in the sense that*

$$\pi(X \times Y) = \pi X \times \pi Y. \qquad (4.2)$$

Notice that (4.2) is subtler than it looks. For on the left-hand side $X \times Y$ is the *topological* product of topological spaces, while on the right-hand side $\pi X \times \pi Y$ is the *direct* product of groups.

Now there is a construction dual, in a precise sense, to that of product, which we call the *coproduct*. Unlike the product, this takes different forms in different categories — in the category of sets it is the disjoint union, as also in the category of topological spaces; in the category of pointed sets (or the category of pointed topological spaces) it is the disjoint union **with base points identified**; in the category of groups it is the free product, . . . However, the formalization proceeds as for the product. Thus, the triple $(Q; q_A, q_B)$ is the *coproduct* of A and B if

(i) $q_A: A \longrightarrow Q$, $q_B: B \longrightarrow Q$ are morphisms; and

(ii)   given any object X and morphisms $g_A: A \longrightarrow X$, $g_B: B \longrightarrow X$, there exists a unique morphism $g: Q \longrightarrow X$ such that $g q_A = g_A$, $g q_B = g_B$.

The appropriate diagram is

$$(4.1')$$

The reader will now see, by comparing $(4.1')$ with (4.1), in what sense the coproduct is *dual* to the product; we have merely *reversed arrows!*

We will write $A * B$ for Q. It now follows from the development in Section 3 that the functor π from polyhedra to groups is coproduct-preserving.[5] That is,

THEOREM Q $\pi : \boxed{Polyhedra} \longrightarrow \boxed{Groups}$ *is coproduct-preserving in the sense that*
$$\pi(X * Y) = \pi X * \pi Y. \qquad (4.2')$$

Again, we draw attention to the fact that $(4.2')$ asserts that the fundamental group of the disjoint union of two polyhedra X, Y, (see Figure 1) with base points identified, is the free product of their fundamental groups.

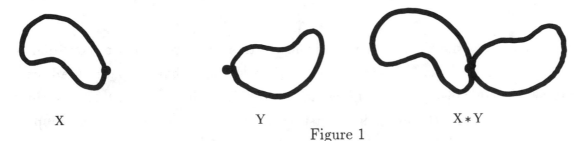

$$\begin{array}{ccc} X & Y & X * Y \end{array}$$

Figure 1

We claim that we have now said quite enough to justify our claim that π is a beautiful functor. On the one hand, as shown in Section 2, it tells us something very important about topological spaces; next, it enables us, as shown in Section 3, in a certain sense to embed group theory in the study of simplicial complexes; and Theorems P and Q show that it preserves some of the most important structural properties of the categories it links. Notice that it would become less perfect if it did much more! If, to take the extreme case, it were to be bijective on objects and morphisms, it would be useless, since it would mean that the two mathematical theories it linked are equivalent.

---

[5] In fact, π is *not* coproduct-preserving over the entire category of pointed spaces; one needs to invoke a 'local' condition on a neighbourhood of the base point to ensure the property. See [2].

Let me close this essay by pointing to two very important consequences of Theorems P and Q. We say that a space X is an *H-space* (or *Hopf space*) if it admits a continuous multiplication with the base point as two-sided identity. Thus an H-space is a substantial generalization of a **topological group**, since neither associativity nor (continuous) inverse are demanded. Now there is a classical theorem that the fundamental group of a topological group is commutative, and this theorem has received some very sophisticated proofs (see, eg., Pontryagin's proof, which is essentially the proof of Theorem 6.9.10 of [2]). Now we may talk of an *H-structure* in *any* category; thus the object X admits an H-structure if there is a morphism $\mu: X \times X \longrightarrow X$ with two-sided identity. Moreover, it is a trivial consequence of Theorem P that

$$\pi \text{ preserves } H\text{-structures.} \tag{4.3}$$

Thus, to show that the fundamental group of an H-space is commutative, we only have to prove

THEOREM H   *Let $\mu: G \times G \longrightarrow G$ be a homomorphism which is an H-structure (in the category of groups). Then G is commutative.*

PROOF   Since μ has a 2-sided identity, e say, we have

$$\mu(g, e) = \mu(e, g) = g, \, \forall g \in G.$$

Thus $\mu(g, h) = \mu(g, e)\, \mu(e, h) = gh$; but also $\mu(g, h) = \mu(e, h)\, \mu(g, e) = hg$. Thus $gh = hg$. Notice that we have proved more, namely, that μ must simply be the group-multiplication.

We turn, finally, to the dual. There is an interesting and important notion dual to that of an H-space, namely, that of an H'-space. Here we ask that a space X admit a continuous **comultiplication**, meaning a map $m: X \longrightarrow X * X$, with base point as two-sided **coidentity**. We must explain these terms. First we only ask for these properties up to homotopy, for a reason we will give shortly. Second we speak of the base point as coidentity in the following sense. The definition $(4.1')$ or the actual construction of the coproduct $X * Y$ (Figure 1) shows that there are projections $r_1: X * Y \longrightarrow X$, $r_2: X * Y \longrightarrow Y$, such that $r_1 q_X = Id_X$, $r_1 q_Y = \text{const.}$, $r_2 q_X = \text{const.}$, $r_2 q_Y = Id_Y$. Then we say that m is an H'-structure if $r_1 m \simeq Id_X$, $r_2 m \simeq Id_X$. Notice that it would be absurd to ask for equality instead of homotopy, since then the only spaces admitting such a comultiplication would be one-point spaces! However, this creates no difficulty for us since π, of course, converts homotopy into equality (property (iii) in the introduction).

Now there are very interesting H'-spaces, the so-called suspension spaces (see, eg., [2, p. 336]) which include all the spheres (of any dimension); however, once again, we are dealing with a substantial generalization of suspension spaces when we discuss H'-spaces. It is easily deducible from the calculations in Section 3 that the fundamental group of a suspension space is free. However, we may prove that the fundamental group of any H'-space is free by first remarking that it is a trivial consequence of Theorem Q that

$$\pi \text{ preserves } H'\text{-structures} \tag{4.3'}$$

and then proving

THEOREM H$'$ Let $\mu : G \longrightarrow G * G$ be a homomorphism which is an H'-structure (in the category of groups). Then G is free.

PROOF We will only sketch the proof. We first observe that the statement that μ is an H'-structure amounts to the assertion that the diagram

$$
\begin{array}{ccc}
G & \xrightarrow{\;\mu\;} & G * G \\
 & \searrow{\scriptstyle \Delta} & \downarrow{\scriptstyle j} \\
 & & G \times G
\end{array}
\tag{4.4}
$$

commutes. Here j is the obvious projection from the free product to the direct product and Δ is the *diagonal* map $\Delta g = (g, g)$. Now one may prove that, for *any* group G, the subgroup $j^{-1}(\Delta G)$ of $G * G$ is a free group on the free generating set $\{ g' g'', \; g \in G, \; g \neq e \}$, where $g'(g'')$ copies g into the first (second) factor of the free product $G * G$.

Returning to (4.4) one sees that, since $\Delta = j\mu$, it follows

(i) that μ is one-one, since Δ is obviously one-one; and

(ii) that μ maps G into $j^{-1}(\Delta G)$.

Thus μ embeds G as a subgroup of a free group. But we know (see Section 3) that a subgroup of a free group is free, completing the proof of the theorem. I cannot forbear to remark that we used the fundamental group already to show that a subgroup of a free group is free — what a truly wonderful functor the fundamental group is!

REFERENCES

1.   Penrose, Roger, "The role of aesthetics in pure and applied mathematics research," *Institute of Mathematics and its Applications* (1974), p. 266.

2.   Hilton, P. J., and S. Wylie, *Homology Theory, C. U. P.* (reprinted 1965).

Calendars, Pacemakers, and Quasicrystals

Marjorie Senechal
Smith College
Northampton, MA 01063

Introduction

It is an honor to have been invited to participate in this conference in honor of Howard Eves. Although I had never met Howard before this occasion, I have admired his books for many years. *A Survey of Geometry* is a favorite reference, always near my desk. Throughout geometry's lean years in our era of abstraction there have been a few special people who, through their excellent and lively writing, have helped to keep the subject alive, and Howard is one of them.

Today geometry is alive and well, not only within "pure" mathematics but in many fields of application. Galileo wrote that the book of nature is written in the characters of geometry. In this paper I will try to show that some very simple characters play an important role in several branches of contemporary science that at first may appear to have little in common.

Some Mathematical Circles

The venerable cycloid is the path traced by a point on a circle rolling along a straight line. The geometrical characters we will be discussing are two interesting variants on this theme. One is the epicycloid, the path of a point on a circle rolling along the circumference of a second circle, on the outside, and the other is the hypocycloid, the path of a point of a circle rolling along the circumference of a second circle, on the inside. Epicycloids were used by ancient astronomers to describe planetary motion; indeed, epicyclic constructions persisted until Kepler. You can get a computer to draw all kinds of cycloids for you.

Blue Moons and Circle Maps

Some medieval clocks were festivals of circles; they had dials showing not only the hours but also the lunar cycle, the solar cycle, and sometimes other cycles as well. Since these cycles don't quite match up, careful planners are faced with a wealth of calendar problems, such as determining the date of Easter. Mathematicians usually think of calendar problems as arithmetic, but they have interesting geometrical interpretations. Consider "blue moons," for example. A blue moon occurred on January 30, 1991. Despite years of singing the song, "Blue Moon," I don't think I knew what one was until January 29, when a radio announcer explained it: when a second full moon occurs in a single calendar month, it is said to be "blue." Blue moons don't occur very often, hence our expression "once in a blue moon." The reason for their scarcity is the incommensurability of the lunar and solar cycles, and the fact that our calendar months have different lengths.

We can visualize the blue moon phenomenon as a hypocylcoid (or epicycloid). Take two circles C_1 and C_2, whose radii R_1 and R_2 satisfy the following equation.

$$\frac{R_1}{R_2} = \frac{\text{length of lunar cycle}}{\text{length of solar cycle}}$$

We mark a point P on the smaller circle C_1, divide the larger circle C_2 into 12 arcs proportional to the lengths of the months, and then divide the month arcs into days. Next place C_1 inside C_2, so that P coincides with the date January 30. Now let C_1 roll along C_2, and mark the points on C_2 where P touches it. Whenever two of these points lie in a single month arc, the second will be a blue moon (Figure 1).

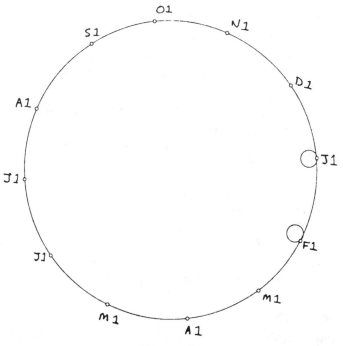

Figure 1
The "Blue Moon" as a Hypocycloid Schematic

Many interesting phenomena can be described by pictures like this. Let's make it much simpler, though. Again we start with two circles C_1 and C_2 of different radii, but this time we divide the circumference C of the larger circle C_2 into two complementary arcs of lengths Δ and $C - \Delta$. Color one arc red, say, and the other green. Mark a point P on C_1 and follow its path as this circle rolls on the inside (or on the outside) of C_2. Sometimes P will land in the red arc, sometimes in the green; in this way we generate an infinite sequence of R's and G's. Different choices of arcs, and different ratios R_1/R_2 give us a family of sequences with some very interesting properties.

Suppose, for example, that the circumference of the larger circle C_2 is 1 and that $\Delta = \frac{1}{2}$. Let the circumference of C_1 be $1/\sqrt{2}$. Mark a starting point 0 on C_2, and measuring clockwise, let the arc Δ from 0 to $\frac{1}{2}$ be red and let the arc 1-Δ be green. Let P coincide with 0 initially. Then the arc from 0 to the point P_1 at which P first returns to C_2 has length $1/\sqrt{2}$. Since $1/\sqrt{2} = .7071...$, P_1 is green. The next time P touches C_1, the rolling circle will have traced out an arc of length $\sqrt{2}$, or $1 + .41421...$ (Figure 2). We can ignore the 1, since that represents a full revolution around the circumference of C_2. Since $.41321... < .5$, the point P_2 is red. Continuing this way, we get the following sequence $\{P_n\}$.

$$P_n = \begin{cases} R, \text{ if } 0 \le n/\sqrt{2} - \left[n/\sqrt{2}\right] < 1/2 \\ G, \text{ otherwise.} \end{cases}$$

([x] is the greatest integer less than or equal to x.) P_n is a function from the integers to the set of two letters $\{R,G\}$; it is a simple dynamical system known as a "circle map." Starting with $n = 0$, the first ten terms of this particular sequence are as follows.

$$R,G,R,R,G,G,R,G,G,R$$

So far, it appears to be irregular. Will it eventually repeat? That could only happen if, for some integers m and n, P_m coincides with P_n , and that in turn could ony happen if $m/\sqrt{2} - n/\sqrt{2} =$ some integer k. but then we would have $(m-n)/k = \sqrt{2}$, which is impossible since $\sqrt{2}$ is irrational. Thus the infinite sequence is nonperiodic.

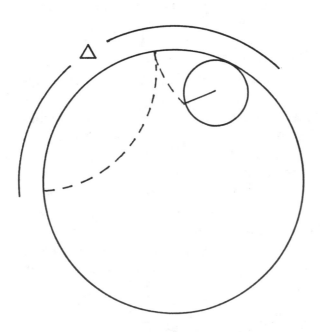

Figure 2
The Hypocycloid as a Circle Map

There are many more interesting questions we can ask about this sequence. What is the relative frequency of R's and G's? From number theory, in particular the theory of uniform distribution modulo one, we learn that in the long run there will be equal numbers of each. This is because $\Delta = 1 - \Delta$; it is independent of the choice of $1/\sqrt{2}$ for the circumference of C_1. When the ratio of the two circumferences is irrational, the ratio of R's to G's will be $\Delta/(1-\Delta)$.

Other interesting questions are: will three or more consecutive R's (or G's) ever occur in this sequence? (It is easy to see that the answer is "no.") Which finite strings of letters do appear, and how often? (This is much harder.) The answers to these questions depend on the particular choice of $1/\sqrt{2}$. We might also ask how closely the relative frequency of R's and G's in large but finite strings of letters approximates the fifty-fifty frequency rate. Such questions may appear to lead us away from geometry into other realms of mathematics, but in fact these other realms enrich our geometrical understanding.

Pacemakers and Quasicrystals

On Sunday, April 21, 1991, a few weeks before the well-publicized Presidential Fibrillation, the Associated Press distributed an article entitled "Tests Find Electric Pulse Signals Heart Attack." The writer explained that researchers had found "...a faint electrical pulse in the heart that preceded ventricular fibrillation" and quoted a researcher as saying, "We saw that whenever the heart is prone to ventricular fibrillation and sudden death, there is always an alternation in the T-wave. It alternates and then the heartbeat goes into a chaotic state."

This was probably no surprise to theorists. What we might call "heartbeat geometry" has been studied in great detail by physiologists, physicians, and mathematicians in recent years, as we learn from *Mathematical Approaches to Cardiac Arrythmias* [2]. There are many causes of ventricular fibrillation, and this is not a lecture on heart disease. But one of these causes, called "ventricular parasystole," might also be called the "blue moon syndrome."

In ventricular parasystole, a second "ectopic" ventricular node develops in the heart and produces a beat that competes with the normal one. In a 1986 paper [3], Glass, Goldberger, and Bélair presented a model for parasystole based on a circle map like the one discussed above. They have since modified some of the features of this simple model (see [2]), but their original version is still of considerable interest, at least to us. Assume that the normal (sinus) beat occurs at time intervals of length one, and that a second ectopic beat occurs at intervals of length ρ. After each normal beat, there is a "refractory time" or time interval Δ. When an ectopic beat occurs in Δ, it is blocked. But when an ectopic beat occurs outside the refractory time, it is conducted and *blocks the subsequent sinus beat* (Figure 3). In other words, the pattern of normal beats is specified by the following function $f(n)$:

$$f(n) = \begin{cases} \text{normal beat,} & \text{if } \rho - [\rho] \in \Delta, \\ \text{no normal beat,} & \text{otherwise} \end{cases}.$$

When ρ is irrational the pattern of normal beats will be irregular. A closer study of this pattern led to a deeper understanding of this type of arrythmia.

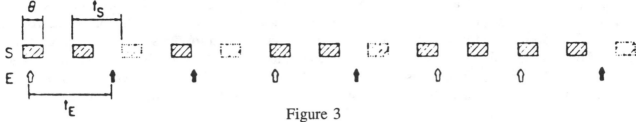

Figure 3

A Simple Circle Map Model for Parasystole (from [3])

In the course of their research, these physiologists consulted the number theorist Michel Waldschmidt in Paris. A few years later, at a conference on Number Theory and Physics, Waldschmidt learned that the same model was being used to study quasicrystals [4]!

What are quasicrystals, and what could they possibly have to do with heartbeats? Quasicrystals are crystals whose symmetry is "forbidden" by the "laws" of crystals. At first this sounds like a contradiction in terms, but let's look more closely.

Until very recently, it was assumed (for good reason) that the atoms in a crystal are always arranged in a periodic array. The smooth polyhedral faces of a crystal were thought to be a consequence of a simple building block structure on the atomic level; the (hypothetical) building blocks, stacked together like cubes to fill all of space, were thought to contain identical groups of atoms or molecules. But if crystals are built this way, then their possible symmetries are severely restricted. To see why, represent each building block by the point at its center. This infinite set of points is called a *three-dimensional lattice*. More precisely, a three-dimensional lattice is the set of endpoints of the linear combinations $a_1\vec{v}_1 + a_2\vec{v}_2 + a_3\vec{v}_3$, where the vectors $\vec{v}_i$ are linearly independent and the a_i are integers. Some two-dimensional lattices are shown in Figure 4.

Figure 4

Some Two-Dimensional Lattices

The distance between any pair of points in a lattice is always greater than or equal to some minimum distance d. Suppose that x and y are two lattice points located at this distance from one another, and assume that a rotation axis passing through x is orthogonal to the line joining x and y. If we rotate about this axis through an angle of $2\pi/k$, y will be moved to another position y' at the same distance d from x; if the rotation is a symmetry operation for the lattice, then y' will also be a lattice point. Now the distance between y and y' must also be greater than or equal to d, so the triangle x,y,y' cannot be acute. We conclude that $k = 2, 3, 4, 5,$ or 6. Two-dimensional lattices for which $k = 2, 3, 4, 6$ are shown in Figure 4. The case $k = 5$ is not shown there, because it is impossible. The points of a lattice have the same surroundings, so if a five-fold axis passes through x, another must pass through y. Now consider the four points x, x', y, y' shown in Figure 5. If the rotations about x and y were four-fold, these four points would form a square; if it were six-fold, x' and y' would coincide. But in this case, five-fold, x' and y' are distinct, and the distance between them is less than d. This contradiction shows that five-fold symmetry is impossible in a two dimensional lattice, and the same argument is valid for three dimensional lattices.

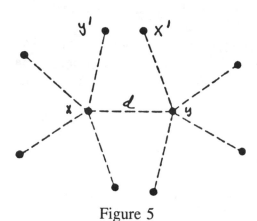

Figure 5
Five-Fold Symmetry is Incompatible with Discreteness

Since the early nineteenth century, the "crystallographic restriction" has been accepted as a law of crystallography. In 1984, however, crystals were discovered that did not obey it! The x-ray patterns they produced clearly showed *five-fold symmetry.* X-ray diffraction patterns are not direct images of atomic structure, but they do provide clues to the structure, and one of those clues is symmetry. What went wrong? The problem is not the argument above, but the hypothesis that all crystal structures are periodic. Evidently, the structure of these crystals is not lattice-like! That is why they were at first called "quasicrystals;" people were reluctant to give up the definition of a crystal as a periodic atomic pattern. Not surprisingly, the discovery of quasicrystals provoked a controversy that raged for several years. Scientists at one extreme claimed that a new state of matter had been discovered; at the other extreme, it was argued that the five-fold symmetry was only apparent and could be explained as a mixture of conventionally crystalline atomic patterns. Today, however, periodicity is no longer considered axiomatic for crystals, and quasicrystals are seen to be one family in a large class of aperiodic crystals [5].

Although nonperiodic, the structures of these crystals are far from random. In particular, they have order sufficient to produce diffraction patterns with bright spots; if the structure were highly disordered, the diffraction pattern would be hazy. What sort of order are we talking about here? This is *the quasicrystal problem*, and it is far from solved. One useful approach is to study nonperiodic tilings [6]; the penrose tilings of the plane (and their analogues in three dimensional space) have received a great deal of attention in this connection. A portion of a Penrose tiling is shown in Figure 6. It is no longer thought that Penrose tilings are models for *real* quasicrystal structure in the same way that building blocks were thought to be models for periodic crystal structure - although the structure of quasicrystals is still not completely understood, experiments have shown that the atoms, wherever they are, are not grouped together in blocks arranged like these. But the vertices of the tiles in a Penrose tiling do produce optical diffraction patterns sufficiently similar to quasicrystal x-ray diffraction patterns to make the analogy worth pursuing. Undoubtedly, we will learn some useful things about orderly nonperiodicity by studying them.

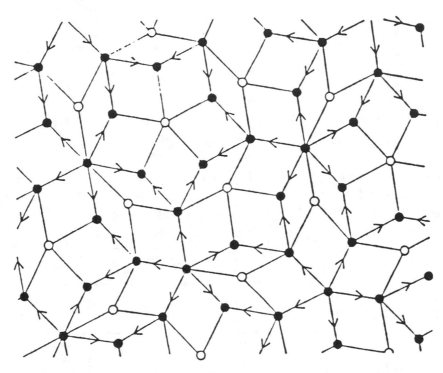

Figure 6
A Portion of a Penrose Tiling of the Plane by Rhombs (from [6])

The best way to begin to understand the Penrose tiles is to cut out copies of them and try to tile the plane. *You must always match a vertex to one of the same color and align the arrows.* The colored vertices and arrowed edges are one expression of a "matching rule" which must be followed to ensure that the tiling is nonperiodic. Tiling the plane in this way is harder than it looks. But after a while you begin to see that the pattern repeats, although not periodically. If you choose any configuration, you will find other copies of it in the pattern, but they will not be equally spaced, or aligned in parallel rows.

Penrose tilings can also be produced by projection from five-dimensional space! Think of a five-dimensional space filled with five-dimensional cubes. Five-dimensional cubes have four-dimensional faces, zero-dimensional vertices, and also facets of one, two, and three dimensions. The Penrose rhombs, whose acute angles are $\pi/5$ and $2\pi/5$, are projections of the two-dimensional facets. A five-dimensional cube has five mutually orthogonal edges meeting at each vertex, and each vertex is joined to a second one on the opposite side of the cube by a body diagonal. If you place a plane at one vertex orthogonal to the body diagonal, and project those five vectors onto that plane, you will get the five-armed star of Figure 7a. The vector sum of two adjacent arms of that star is the fourth vertex of a thick Penrose Rhomb; the vector sum of two nonadjacent edges is the fourth vertex of a thin one (Figure 7b). The Penrose tilings are projections of continuous surfaces of such two-dimensional facets.

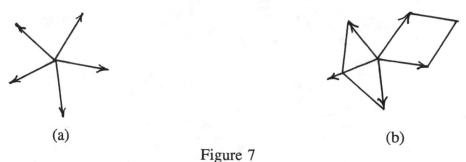

(a)   (b)

Figure 7
(a) An Orthogonal Projection of Five Edges of a Five-Dimensional Cube Onto a Plane
(b) The Edges Define the Penrose Rhombs

Quasicrystals with 8-fold, 10-fold, and 12-fold symmetry have also been discovered in the last few years, as have one dimensional quasicrystals. All of them are being studied by projected tiling models. Let's take a closer look at the one-dimensional case.

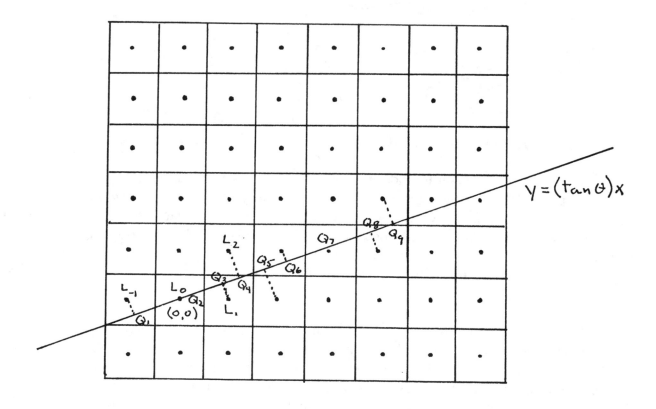

Figure 8
A Projection Model for One-Dimensional Quasicrystals

Consider the ordinary square lattice in the plane, and draw square boxes of edge length 1 with the lattice points at their centers. Now draw a line through the origin with irrational positive slope $\tan \theta$ (Figure 8). Starting with $L_0 = (0,0)$, let $L_0, L_{\pm 1}, L_{\pm 2}, \ldots$ be the lattice points whose boxes are cut by the line. We can join L_i to L_{i+1} by a unit vector, either $\vec{e}_1 = (1,0)$ or $\vec{e}_2 = (0,1)$. The chain of $\vec{e}_1$'s and $\vec{e}_2$'s which link the L_i forms a staircase; it is the analogue of the surfaces mentioned above. When we project the staircase onto the line we get a tiling of the line whose vertices Q_i are the projections of the lattice points L_i. The distance between Q_i and Q_{i+1} will be $r = \cos \theta$ if the corresponding lattice points L_i and L_{i+1} are joined by the vector $\vec{e}_1$ and $g = \sin \theta$ if they are joined by the vector $\vec{e}_2$. Since $\tan \theta$ is irrational, the sequence of r's and g's is nonperiodic. The points Q_i are the vertices of a one-dimensional "quasicrystal".

This sounds somewhat like a circle map. And so it is. Let us complete the circle of ideas. Let the unit vector along $y = x \tan \theta$ be $\vec{u}_1 = (\cos \theta, \sin \theta)$ and let $\vec{u}_2$ be the orthogonal unit vector $(-\sin \theta, \cos \theta)$. Let $L_i = (m,n)$ be the lattice point corresponding to Q_i. The component of L_i in the $\vec{u}_2$ direction is $c_i = -m \sin \theta + n \sin \theta$, and the $\vec{u}_2$ component of L_{i+1} will be as follows.

$$\begin{cases} c_i + \cos \theta, & \text{if } L_{i+1} - L_i = \vec{e}_2, \\ c_i - \sin \theta, & \text{otherwise.} \end{cases}$$

As Figure 9 shows, the projection of L_i onto $\vec{u}_2$ always lies in a fixed interval I of length $\cos \theta + \sin \theta$. We divide I into two subintervals, I_1 of length $\sin \theta$, and I_2 of length $\cos \theta$. Color I_1 red and I_2 green. If L_i projects into the red interval, we will have $c_{i+1} = c_i + \cos \theta$; if L_i projects into the green interval, then $c_{i+1} = c_i - \sin \theta$.

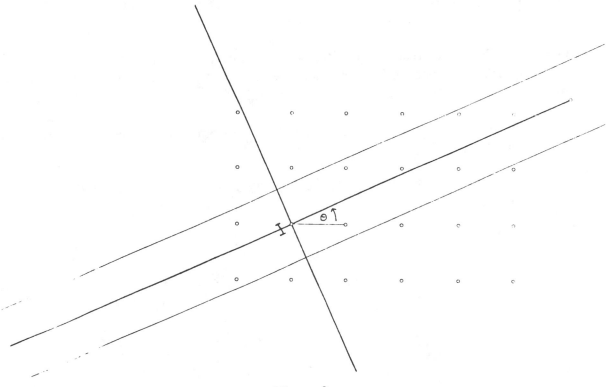

Figure 9
The Projection Model is Specified by a Circle Map

So far we have seen no circles, and circle maps involve two of them, a fixed circle and a rolling one. We get our fixed circle by identifying the endpoints of the interval I. The rolling circle is a circle of circumference $\cos \theta$; notice that $c_i - \sin \theta = c_i + \cos \theta - (\cos \theta + \sin \theta)$. That is, the sequence of $\vec{u}_2$ components of the points L_i is the sequence $n \cos \theta \pmod{\cos \theta + \sin \theta}$, and it is this sequence that specifies the sequence of r's and g's.

Conversely, every circle map can be represented as a projection. The questions we asked earlier about circle maps (What is the relative frequency of r's and g's? How many consecutive r's - or g's - can occur in this sequence? What other finite strings of letters appear, and how often? How closely does the ratio of r's to g's in large but finite strings of letters approximate the frequency rates?) can be studied in the projection model. They are crucial to understanding the mathematical and physical properties of quasicrystals.

(I would like to thank Michel Waldschmidt for bringing reference [3] to my attention.)

References

[1]   Wagon, S., *Mathematica in Action*, Freeman, NY, 1990.

[2]   Jalife, J., ed., "Mathematical Approaches to Cardiac Arrythmias," *Annals of the New York Academy of Sciences*, Vol. 591, 1990.

[3]   Glass, L., A. Goldberger, and J. Belair, "Dynamics of pure parasystole," *American Journal of Physiology*, Vol. 251, H841-H847, 1986.

[4]   Godrèche, C., "Types of order and diffraction spectra for the line," in *Number Theory and Physics*, Proceedings of the Winter School, Les Housches, France, March 7-16, 1989, edited by J. M. Luck, P. Moussa, and M. Waldschmidt, Springer-Verlag, Berlin-Heidelberg, 1990.

[5]   Senechal, M. and J. Taylor, "Quasicrystals: the view from Les Houches," *The Mathematical Intelligencer*, Vol. 12, No. 2, 1990, 54-64.

[6]   Grünbaum, B. and G. S. Shepard, *Tilings and Patterns*, Freeman, NY, 1987.

Two Families of Cubics Associated with a Triangle

Peter Yff
Ball State University
Muncie, IN 47306

1. Introduction

In the cartesian plane a cubic curve, or cubic, is the locus of points (x, y) satisfying a polynomial equation of degree three. If homogeneous coordinates, such as trilinear coordinates, are used in the plane, a cubic is the locus of points (x_1, x_2, x_3) satisfying a homogeneous equation of degree three. (More will be said about trilinear coordinates in the next section.) J. E. A. Steggall [4] described the following cubics, expressed in trilinear coordinates.

$$(1) \qquad \sum a_2 a_3 x_1 \left(x_2^2 - x_3^2 \right) = 0,$$

$$(2) \qquad \sum (\cos \alpha_1 - \cos \alpha_2 \cos \alpha_3) x_1 \left(x_2^2 - x_3^2 \right) = 0,$$

$$(3) \qquad \sum (\cos \alpha_1) x_1 \left(a_2^2 x_2^2 - a_3^2 x_3^2 \right) = 0.$$

Here $a_i \ (i = 1,2,3)$ is the length of the side opposite A_i in the triangle $A_1 A_2 A_3$ and α_i is the measure of angle A_i. The summation sign assumes cyclic permutation of subscripts modulo 3. For example, equation (1) means $a_2 a_3 x_1 \left(x_2^2 - x_3^2 \right) + a_3 a_1 x_2 \left(x_3^2 - x_1^2 \right) + a_1 a_2 x_3 \left(x_1^2 - x_2^2 \right) = 0$.

F. G. Taylor [5] rediscovered cubics (1) and (2). L. Miller [3] arrived at (2), and possibly (3), by a method which yields both cubics simultaneously. In this paper it will be shown that (1) and (2) are special cases of an infinite family of cubics, each passing through seven fixed points. Cubic (3) belongs to another infinite family with seven points in common. Each of these families may be defined by a single condition expressed in elementary terms.

2. Isogonal Cubics

Trilinear coordinates with respect to triangle $A_1 A_2 A_3$ will be used throughout. If d_1, d_2, d_3 are the directed distances of a point P from lines $A_2 A_3, A_3 A_1, A_1 A_2$ respectively, then any triple $(kd_1, kd_2, kd_3), \ (k \neq 0)$, may be taken as trilinear coordinates of P. The sense of a directed distance is determined by the assumption that all distances are positive from any point inside the triangle. For a point outside the triangle but within angle $A_2 A_1 A_3$ the distance d_1 to $A_2 A_3$ is considered as being negative. This rule applies similarly to the other coordinates. Whenever a point crosses a line A_iA_j the corresponding coordinate changes sign. For example, $(1, 0, 0)$ represents A_1, while $(1, 1, 1)$ represents the incenter I (center of the inscribed circle). Cubics (1) and (2) have nine points in common. These are the vertices A_1, A_2, A_3, the incenter I, the excenters $I_1(-1,1,1), I_2(1,-1,1), I_3(1,1,-1)$, the circumcenter O $(\cos\alpha_1, \cos\alpha_2, \cos\alpha_3)$ (center of the circumscribed circle), and the orthocenter H $(\sec\alpha_1, \sec\alpha_2, \sec\alpha_3)$ (intersection point of the altitudes). In addition, (1) goes through the centroid G $(a_1^{-1}, a_2^{-1}, a_3^{-1})$ (intersection of the medians), the Lemoine point K (a_1, a_2, a_3) (intersection of the symmedians), and the midpoints $(0, a_3, a_2), (a_3, 0, a_1), (a_2, a_1, 0)$ of the sides. On the other hand, (2) goes through the point of de Longchamps L $(\cos\alpha_1 - \cos\alpha_2\cos\alpha_3, \cos\alpha_2 - \cos\alpha_3\cos\alpha_1, \cos\alpha_3 - \cos\alpha_1\cos\alpha_2,)$, its isogonal conjugate L', and the feet of the cevians through L.

In order to clarify some of the above, some definitions are in order. Any line from a vertex to the opposite side, such as a median or an altitude, is called a cevian. In some cases the side may be extended, with the cevian being outside the triangle. The foot of the cevian is the point at which it meets the opposite side. If two cevians from A_1 are situated so that the angle between them is bisected by the bisector of angle $A_2A_1A_3$, the cevians are called isogonal conjugates of each other. If cevians from the three vertices are concurrent at P, it is known [3] that their isogonal conjugates are concurrent at a point P', which is called the isogonal conjugate of P. If the trilinear coordinates of P are (x_1, x_2, x_3), the coordinates of P' are $(x_1^{-1}, x_2^{-1}, x_3^{-1})$. For example, K is the isogonal conjugate of G, and each symmedian is the isogonal conjugate of the corresponding median.

The points O, G, and H are on one line, called the Euler line of the triangle [3]. Moreover, G is always between O and H with $HG = 2GO$. The de Longchamps point L is on the same line, with O the midpoint of HL.

Now let $C(c_1, c_2, c_3)$ be a fixed point, and consider the locus of $P(x_1, x_2, x_3)$, constrained to move so that P and its isogonal conjugate P' are collinear with C. This locus satisfies the following equation.

$$\begin{vmatrix} x_1 & x_2 & x_3 \\ x_1^{-1} & x_2^{-1} & x_3^{-1} \\ c_1 & c_2 & c_3 \end{vmatrix} = 0$$

After multiplication by $x_1x_2x_3$, this becomes

(4) $\sum c_1 x_1 (x_2^2 - x_3^2) = 0.$

This cubic goes through the seven points $A_1, A_2, A_3, I, I_1, I_2,$ and I_3, as well as C, its isogonal conjugate C' and the feet of the cevians through C. The last three points have coordinates $(0, c_2, c_3), (c_1, 0, c_3), (c_1, c_2, 0)$ respectively. For each of the twelve points mentioned, the coordinates can be shown to satisfy (4).

As C moves over the plane, (4) represents a two-parameter family of cubics. Each cubic is isogonally self-conjugate, in the sense that for any point P on the curve, its isogonal conjugate P' (if it exists) also belongs to the curve. Therefore, this will be called the family of isogonal cubics associated with triangle $A_1A_2A_3$.

If C is on a line joining two of the isogonal self-conjugate points I,I_1,I_2,I_3, its cubic degenerates into the union of that line and a conic. If C coincides with one of these points, its cubic degenerates into three lines. These lines are the internal angle bisectors of the triangle (if $C=I$), or one internal and two external bisectors through C (if C is an excenter). No further attention will be paid to these cases.

Cubics may be classified as rational or irrational (see Hilton [1]). In the former case the variables may be expressed as rational functions of a parameter. This kind of cubic is also called singular, because it has a double point, which may be a node, a cusp, or an isolated point. The isogonal cubics are irrational, so there is never a double point. Also every irrational cubic consists of two branches; an "odd" branch which has a real point at infinity, and an "even" branch resembling a conic section. The even branch may be called elliptic, parabolic, or hyperbolic, according as it is a bounded oval, a curve with one point at infinity, or a curve with two distinct points at infinity. It may even be imaginary, but this does not happen if the cubic is isogonal, as will now be shown.

For each isogonal cubic the tangents at I,I_1,I_2,I_3 are concurrent at C. The reason is that each of these four points is isogonally self-conjugate. If IC were to intersect the cubic at another point P, it would also have to meet it at P', the isogonal conjugate of P. This is impossible because a line and a cubic have at most three points in common. Thus there are four real tangents to the cubic from C, so the even branch must be real. It may also be shown that the tangents at A_1,A_2,A_3, and C are concurrent at C', the isogonal conjugate of C.

The tangent line to (4) at any point (y_1,y_2,y_3) may be found by assuming its equation to be $(ty_2-y_3)x_1-ty_1x_2+y_1x_3=0$ for some value of t (or $y_2x_1=y_1x_2$ if t goes to infinity). Solving this simultaneously with (4), assuming a double solution at (y_1,y_2,y_3), gives the value of t and the folllowing equation of the tangent line.

$$(5) \qquad \sum \left[c_1\left(y_2^2-y_3^2\right) - 2y_1(c_2y_2-c_3y_3)\right]x_1 = 0$$

Several interesting special cases are obtained when C is collinear with H,G, and O, that is, when C is on the Euler line. First, let $u_1x_2x_3 + u_2x_3x_1 + u_3x_1x_2 = 0$ be the equation of a circumconic of $A_1A_2A_3$ (any conic through the vertices). Under what conditions are the normals to the conic at the vertices concurrent? The line normal to the conic at A_1 is given by

$$(u_2 - u_3\cos\alpha_1)x_2 - (u_3 - u_2\cos\alpha_1)x_3 = 0,$$

and the normals at A_2 and A_3 are obtained by cyclic permutation of subscripts. If the three normals are concurrent, then

$$(u_2 - u_3 \cos\alpha_1)(u_3 - u_1 \cos\alpha_2)(u_1 - u_2 \cos\alpha_3) = (u_3 - u_2 \cos\alpha_1)(u_1 - u_3 \cos\alpha_2)(u_2 - u_1 \cos\alpha_3),$$

which reduces to

$$\sum(\cos\alpha_1 + \cos\alpha_2 \cos\alpha_3)u_1(u_2^2 - u_3^2) = 0, \text{ or } \sum(\sin\alpha_2 \sin\alpha_3)u_1(u_2^2 - u_3^2) = 0,$$

showing that the locus of (u_1, u_2, u_3), regarded as a point, is (1), the isogonal cubic associated with G.

Since (u_1, u_2, u_3) has been considered merely as an ordered triple, it should also be noted that it represents a point associated with the circumconic. The feet of the cevians through (u_1, u_2, u_3) are $(0, u_2, u_3)$, $(u_1, 0, u_3)$, and $(u_1, u_2, 0)$. Now the harmonic conjugate of $(0, u_2, u_3)$ with respect to A_2 and A_3 is $(0, u_2, -u_3)$, and the line through this point and A_1 is tangent to the conic at A_1. (If P, Q, and R are points on a line, the harmonic conjugate of R with respect to P and Q is the point S on the same line such that $PR / RQ + PS / SQ = 0$, in which directed distances are used.) Similarly the harmonic conjugates of $(u_1, 0, u_3)$ and $(u_1, u_2, 0)$, may be found, with their corresponding tangent lines. Furthermore, the three harmonic conjugates are on the line $\sum u_i^{-1} x_i = 0$, which is called the trilinear polar of the point (u_1, u_2, u_3).

While (u_1, u_2, u_3) moves along cubic (1), the conic changes, but the three normals remain concurrent. The locus of their common point may be shown, after some calculation, to be cubic (2). This is the isogonal cubic determined by the de Longchamps point L, as mentioned earlier.

Miller [3] shows that the perpendicular bisectors of the sides of the triangle are the asymptotes of (2). In addition, this cubic is symmetrical with respect to a half turn about O (see Figure 1). One advantage of using trilinear coordinates (or any homogeneous coordinate system) is that points at infinity can be treated as easily as other points. In this system all points at infinity lie on the line whose equation is $\sum a_i x_i = 0$. This line intersects (2) at the points $(-1, \cos\alpha_3, \cos\alpha_2), (\cos\alpha_3, -1, \cos\alpha_1)$, and $(\cos\alpha_2, \cos\alpha_1, -1)$ The first of these points is on every line perpendicular to $A_2 A_3$, in particular on the perpendicular bisector of $A_2 A_3$: $\sin(\alpha_2 - \alpha_3)x_1 + (\sin\alpha_2)x_2 - (\sin\alpha_3)x_3 = 0$. Simultaneous solution of this equation with (2) reveals a double solution at infinity. Hence the line is tangent to the cubic at infinity and is therefore an asymptote. Similarly perpendicular bisectors of the other sides are also asymptotes. However, this case is unusual in that the asymptotes of an isogonal cubic are seldom elegantly expressed; finding them involves solving a cubic equation.

Another remarkable case is the isogonal cubic associated with the circumcenter O. Its equation is $\sum(\cos\alpha_1)x_1(x_2^2 - x_3^2) = 0$. In solving for points at infinity, it is found that the resulting cubic is reducible, giving the coordinates of one point as

$$\left(-\cos\frac{\alpha_2 - \alpha_3}{3}, \cos\frac{\alpha_2 + 2\alpha_3}{3}, \cos\frac{2\alpha_2 + \alpha_3}{3}\right)$$

The other two points at infinity are obtained by cyclic permutation of subscripts. Each of these three points may be shown to lie on an axis of symmetry of Steiner's deltoid (see Zwikker [7]). This is the envelope of the system of pedal lines (Simson lines) of $A_1 A_2 A_3$ and is a hypocycloid of three cusps.

Since the deltoid has all of the symmetries of an equilateral triangle, the aymptotes of the cubic make angles of 60° at their intersection points. In fact, it may be shown that they are concurrent at the centroid G (see Figure 2). Again, this elegant result is an exceptional case, because L and O are the only points on the Euler line whose isogonal cubics have three concurrent asymptotes.

The isogonal cubic of O also appears in the following context. Let R be a point such that the sum of angles $RA_2A_3, RA_3A_1,$ and RA_1A_2 is a right angle. The locus of R is again the cubic associated with O.

In each case considered until now the even branch of the cubic has been hyperbolic. That is, it meets the line at infinity at two distinct points. Now the point C will be taken at infinity, in which case all the lines through C are parallel. Almost every line through C, including the line at infinity, contains a pair of isogonal conjugate points on the cubic. However, the isogonal conjugate points at infinity have imaginary coordinates $(1, -\cos\alpha_3 \pm i\sin\alpha_3, -\cos\alpha_2 \mp i\sin\alpha_2)$. This means that C is the only real point at infinity on the cubic, and the even branch is therefore elliptic. An example is that of $C = E$ on the Euler line, with the coordinates

$$(\cos\alpha_1 - 2\cos\alpha_2\cos\alpha_3, \cos\alpha_2 - 2\cos\alpha_3\cos\alpha_1, \cos\alpha_3 - 2\cos\alpha_1\cos\alpha_2).$$

Since the tangent line at E goes through its isogonal conjugate E' (a point on the circumcircle), the equation of asymptote EE' may be readily calculated. The even branch of this isogonal cubic is a bounded oval (see Figure 3).

While taking C at infinity is sufficient to obtain a cubic with an elliptic even branch, it is not necessary. (Counterexamples have been found but will not be shown here.) Most likely there are also isogonal cubics with parabolic even branches, but computation has not been completed.

One important concept not to be overlooked is that of the polar conic of a point with respect to a cubic. In general, a line through a point P on a cubic intersects the curve in two other points. Whether they be real or imaginary, the harmonic conjugate of P with respect to these points is real, and the locus of the conjugate points of P is a conic, called the polar conic of P with respect to the given cubic ([1], 88-90). In particular, the polar conic of C with respect to cubic (4) must pass through each of the isogonal self-conjugate points $(1, \pm1, \pm1)$ as well as C itself. With these five points the equation of the conic can be shown to be $\sum (c_2^2 - c_3^2)x_1^2 = 0$. Since the incenter and the excenters form an orthocentric system, each being the orthocenter of the triangle formed by the others [2], it is known [7] that any conic passing through them is a rectangular hyperbola. Thus the polar conic of C is a special case. It is also known that the locus of centers of all such hyperbolas is the common nine-point circle of the four triangles determined by the points of the orthocentric system taken three at a time. The nine point circle of a triangle is determined by the feet of its altitudes. It also passes through the midpoints of the sides and the midpoints of the segments joining the orthocenter to the vertices. In the present case the nine-point circle of I, I_1, I_2, I_3 is the circumcircle of $A_1A_2A_3$. Therefore the polar conic of C with respect to its isogonal cubic is the rectangular hyperbola through C, the incenter, and the excenters, and its center is on the circumcircle. By regarding the center of the conic as the pole of the line at infinity with respect to the conic, its coordinates are found to be

$$\left(\frac{a_1}{c_2^2 - c_3^2}, \frac{a_2}{c_3^2 - c_1^2}, \frac{a_3}{c_1^2 - c_2^2}\right).$$

3. Isotomic cubics

Cubic (3) is clearly not isogonal, but it will now be shown to belong to a related family of cubics. The isotomic conjugate of $P(x_1, x_2, x_3)$ is the point $P^*(a_1^{-2}x_1^{-1}, a_2^{-2}x_2^{-1}, a_3^{-2}x_3^{-1})$. If P and P^* are always collinear with C, the equation of their locus is

(6) $\sum a_1^2 c_1 x_1 \left(a_2^2 x_2^2 - a_3^2 x_3^2\right) = 0,$

which is obtained by the same method used for (4).

Cubic (6) goes through A_1, A_2, A_3, the centroid G, and the exmedian points $G_1\left(-a_1^{-1}, a_2^{-1}, a_3^{-1}\right)$, $G_2\left(a_1^{-1}, -a_2^{-1}, a_3^{-1}\right)$, $G_3\left(a_1^{-1}, a_2^{-1}, -a_3^{-1}\right)$, as well as C, C^*, and the feet of the cevians through C. (The exmedian points are such that each of them is the vertex of a parallelogram whose other vertices are A_1, A_2, A_3. In other words, A_1, A_2, A_3 are the respective midpoints of $G_2 G_3, G_3 G_1, G_1 G_2$.) Each of the above twelve points has coordinates that may be shown to satisfy (6).

Since (6) is isotomically invariant, in the sense that if P is any point on the cubic, its isotomic conjugate P^* (if it exists) also lies on the cubic, (6) will be called an isotomic cubic. It has properties analogous to those of isogonal cubics. It is irrational and has no double point. The tangents to (6) at A_1, A_2, A_3, and C are concurrent at C^*. The tangents at the isotomic self-conjugate points, G, G_1, G_2, G_3, concur at C. If C is on any line joining two self-conjugate points, (6) degenerates into the union of that line and a conic. If it lies on two such lines, then it lies on three (coinciding with one of the self-conjugate points), and the three lines form the degenerate cubic (6). Only nondegenerate cubics will be considered here.

Comparison of (3) with (6) shows that the former is an isotomic cubic with $a_i^2 c_i \sim \cos \alpha_i$ $(i = 1, 2, 3)$. Thus $c_i \sim a_i^{-2} \cos \alpha_i$, and C is the isotomic conjugate of the orthocenter H $(\sec \alpha_1, \sec \alpha_2, \sec \alpha_3)$. This cubic appears simultaneously with (2) in a problem considered by Miller [3]. Let Q_1, Q_2, and Q_3 be the feet of the perpendiculars from P to the corresponding sides of the triangle. The problem is to determine the locus of P if the three cevians $A_i Q_i$ are concurrent at a point Q. It is proved that this locus is the isogonal cubic (2), but Miller does not deal with the natural question of what the locus of Q might be, except in some degenerate cases.

To prove that (3) is the locus of Q, let P have coordinates (u_1, u_2, u_3) which satisfy (2). The coordinates of Q_1 can be found by elementary geometry to be $(0, u_2 + u_1 \cos \alpha_3, u_3 + u_1 \cos \alpha_2)$. The equation of $A_1 Q_1$ is $(u_3 + u_1 \cos \alpha_2) x_2 - (u_2 + u_1 \cos \alpha_3) x_3 = 0$, and the equations of the other cevians through Q can be written by cyclic permutation of subscripts. Solving two of these equations simultaneously gives the coordinates of Q as

$$[(u_3 + u_1 \cos\alpha_2)(u_1 + u_2 \cos\alpha_3), (u_2 + u_1 \cos\alpha_3)(u_3 + u_2 \cos\alpha_1), (u_3 + u_1 \cos\alpha_2)(u_3 + u_2 \cos\alpha_1)],$$

with the condition that P is on (2). It may then be verified that the coordinates of Q sastisfy (3) (see Figure 4).

An interesting fact that Miller does not mention is that the line PQ always goes through L. Since L is the determining point of (2), the isogonal conjugate P' is also on this line. The collinearity of these four points may be deserving of study.

Not many examples of isotomic cubics have been investigated, and none has been found with special properties such as three concurrent asymptotes. However, two more examples will be given. The first is presented by a definition analogous to that of R, whose locus is the isogonal cubic of O, as already noted. In this case let T_1, T_2, T_3 be the feet of the cevians through a point T on the respective lines A_2A_3, A_3A_1, A_1A_2. If the sum of the directed lengths A_2T_1, A_3T_2, A_1T_3 (the positive sense on the perimeter being that from A_1 to A_2 to A_3 to A_1) is half the perimeter of the triangle, the equation of the locus of T is $\sum a_1(a_2 + a_3 - a_1)x_1(a_2^2 x_2^2 - a_3^2 x_3^2) = 0$. Comparison with (6) shows that $c_1 \sim \frac{a_2 + a_3 - a_1}{a_1}$, which by the Law of Sines and some identities is proportional to $\csc^2 \frac{\alpha_1}{2}$. The values of c_2 and c_3 are analogous, and the coordinates of C are those of the Nagel point of $A_1A_2A_3$. (One way to define the Nagel point is that it is the point of concurrence of the three cevians A_iN_i, where N_i is the point halfway around the perimeter from A_i.) In Figure 5, N^* represents the Gergonne point, which is the isotomic conjugate of N. It may also be defined as the point of concurrence of the cevians from the vertices to the points where the inscribed circle is tangent to the sides of $A_1A_2A_3$. Incidentally, the coefficient $a_1(a_2 + a_3 - a_1)$ in the equation of the cubic could have been replaced by $\cot \frac{\alpha_1}{2}$.

As a final example C will be taken at infinity. Consider $C = V$, with coordinates $(a_2 - a_3, a_3 - a_1, a_1 - a_2)$. As was proved for isogonal cubics with C at infinity, the even branch of this isotomic cubic is elliptic. Hence there is only one asymptote VV^*. The isotomic conjugate of V, V^*, is on the circumscribed Steiner ellipse of the triangle $\sum a_i^{-1} x_i^{-1} = 0$, whose center is the centroid and whose tangent line at each vertex is parallel to the opposite side. Two notable points on this cubic are the isotomic conjugates U and U^*, defined [6] as analogous to the Brocard points of a triangle. The first Brocard point Ω is defined by the property that angles ΩA_2A_3, ΩA_3A_1 and ΩA_1A_2 are equal, while the second Brocard point Ω' is defined by the equality of angles $\Omega' A_3A_2$, $\Omega' A_1A_3$, and $\Omega' A_2A_1$. Analogously, U is defined as the point with cevians A_iU_i $(i = 1, 2, 3)$ such that $A_2U_1 = A_3U_2 = A_1U_3$, while U^* is the point with cevians $A_iU_i^*$ such that $U_1^*A_3 = U_2^*A_1 = U_3^*A_2$.

The coordinates of U and U^* are expressed in terms of an irreducible cubic, but V at infinity is the unique point on line UU^* with "rational" coordinates. (See Figure 6.)

References

1. Hilton, H., *Plane Algebraic Curves,* Oxford University Press, London, 1932.

2. Johnson, R.A., *Advanced Euclidean Geometry*, Dover, New York, 1960.

3. Miller, L.H., *College Geometry*, Appleton-Century-Crofts, New York, 1957.

4. Steggall, J.E.A., "A Certain Cubic Connected with the Triangle", *Proc. Edinburgh Math. Soc.* VII (1889), 66-67.

5. Taylor, F.G., "Two Remarkable Cubics Associated with a Triangle", *Proc. Edinburgh Math. Soc.* XXXIII (1915), 70-84.

6. Yff, P., "An Analogue of the Brocard Points", *Amer. Math. Monthly* 70 (1963), 495-501.

7. Zwikker, C., *The Advanced Geometry of Plane Curves and Their Applications*, Dover, New York, 1963.

Appendix - Some graphs of cubics

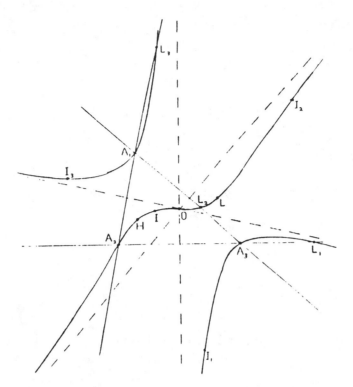

Figure 1 - Isogonal cubic of L

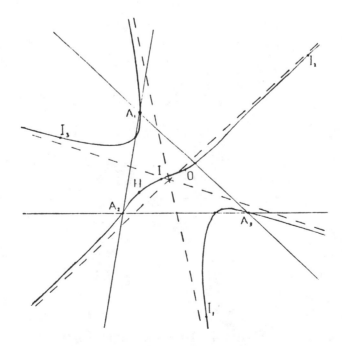

Figure 2 - Isogonal cubic of O

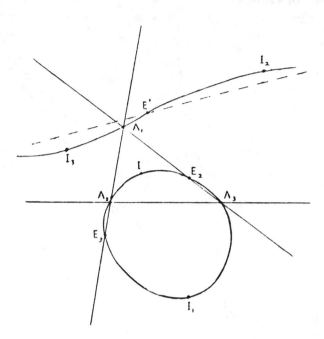

Figure 3 - Isogonal cubic of *E*

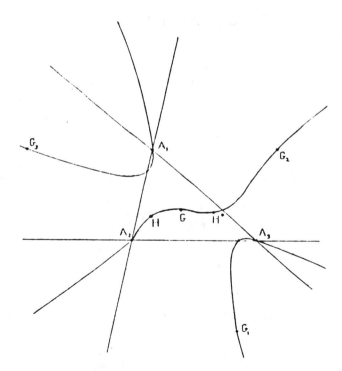

Figure 4 - Isotomic cubic of *H**

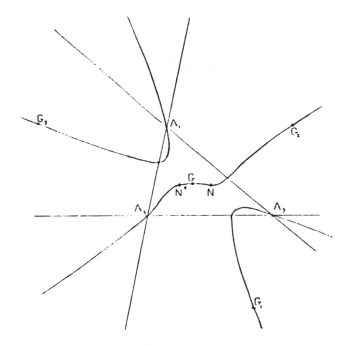

Figure 5 - Isotomic cubic of N

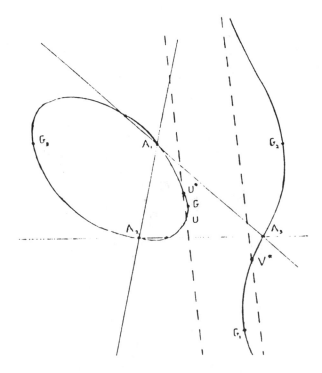

Figure 6 - Isotomic cubic of V

Symmetry Through the Ages
Highlights from the History of Regular Polyhedra

Benno Artmann
Darmstadt Technical University
Darmstadt, Germany

From the time of the Greeks, people have been fascinated by the regular polyhedra. They provide us with what well may be the first complete mathematical theory; a general definition together with a complete classification of all the objects satisfying the definition. I will try to present what I think are the most important steps in the development of a subject that goes back to the beginning of theoretical mathematics and is still alive today. Hippasus provided the first significant example and Theaetetus created the mathematical theory. Pacioli revived the subject which had been dormant for about a thousand years. Felix Klein replaced the polyhedra by their symmetry groups and opened vast new areas of research. One path leads into function theory and algebraic geometry, while the other starts with the group of rotations of the dodecahedron and goes on to simple groups.

Outside of mathematics there flows another stream of ideas related to the regular polyhedra. Plato associated them with the elements. From Roman times we have numerous dodecahedra whose purpose remains a mystery. Pacioli, while keeping the contents of his book strictly mathematical, is so enthusiastic about the polyhedra, and especially the golden section, that he may well have inspired the German art historian Zeising in the middle of the last century. Zeising claimed to have found the key to all secrets of beauty in the proportion of the golden section.

The beauty of the regular solids does not reside in their physical appearance; it lies hidden in the realm of mathematical thought. The interplay of the general concept of regularity and its realization in exactly five solids can be grasped only by mathematics. Plato was the first to understand this. Idea and participation lie at the center of his philosophy.

A.   The Essential First Example:
Hippasus and the Dodecahedron, 480 B.C.

Pythagoras lived about 510 B.C. We have very little firsthand knowledge of him. One of his sayings is reported to be "What are the most beautiful Figures? The circle and the sphere!" This shows that he identified the most symmetric with the most beautiful figures. One of his disciples was Hippasus of Metapontum ($\approx$ 480 B.C.). He belonged to a group of Pythagoras' followers who were later called mathematicians. Iamblichus, writing about 300 A.D., noted that, "About Hippasus in particular they report that he belonged to the Pythagoreans, but because he was the first to make public the secret of 'the sphere of the twelve pentagons,' he perished at sea. The fame of the discoverer, however, was his..." The 'sphere of the twelve pentagons' is, of course, the dodecahedron. A scholion to Euclid gives us more information. It says that Euclid's Book XIII [13] is about "...the five figures called Platonic, which, however, do not belong to Plato. Three of these five figures, the cube, pyramid, and dodecahedron, belong to the Pythagoreans, while the octahedron and icosahedron belong to Theaetetus."

From this and other evidence (see the literature at the end of this section) it is generally thought that Hippasus was indeed the first to treat the dodecahedron mathematically. His 'construction' may have been very similar to the one given by Euclid in his proposition XIII, 17. Euclid starts with a cube and adds (parts of) pentagons like the roof of a house (Figure 1 and Figure 2). If the edge d of the cube is given, he must find the edge f of the pentagon. That is, he must find the ratio d/f, now called the 'golden ratio'. If we assume the faces of the dodecahedron to be regular pentagons, then the edge d of the cube will be one of the diagonals of the pentagon.

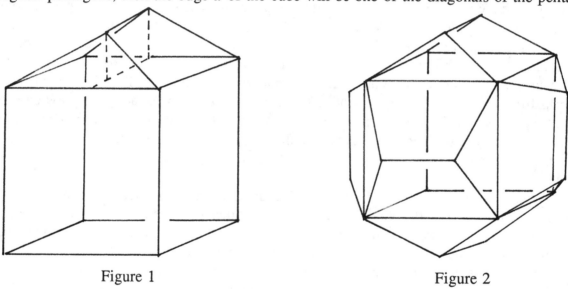

Figure 1   Figure 2

In Figure 3 the similar triangles $\triangle$BCD and $\triangle$CXD imply the proportion $d/f=f/(d-f)$. Hence $d(d-f)=f^2$. Euclid solves this quadratic problem by means of the so-called 'geometric algebra of the Pythagoreans' in proposition VI, 30.

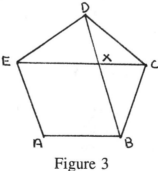

Figure 3

If we replace VI, 30 by one of the constructions at the beginning of Book XIII, then, except for similar triangles, the construction of the dodecahedron in Book XIII would not depend on material in the earlier Books, especially not on the construction of the regular pentagon in Book IV. This may well have happened. I think it likely that discoveries (like that of the golden ratio) involve complicated circumstances, and that simple, direct presentations are the results of later reworking. The books by Burkert [5] and van der Waerden [26] give details about Pythagoras and the Pythagoreans. A detailed study of Hippasus is due to von Fritz [12]. Some additions to this are in Artmann [1]. For the much debated 'geometric algebra', see Artmann [3].

Dodecahedra outside of mathematics

About 390 neolithic carved stone balls of fist size dating from before or about 2000 B.C. have been found in Scotland. All of the five regular solids appear in their decorations, the dodecahedron on one specimen in the Museum of Edinburgh. A scientific description in Ritchie and Ritchie [21] calls them a "...relatively short lived and peculiar Scottish phenomenon." Very good photographs are to be found in Critchlow [10], who is, however, otherwise very speculative. A picture with five 'regular' stone balls is reproduced in *Mathematics Teaching*, Vol. 110 (March, 1985), p. 56.

Dodecahedra made of bronze were popular in Roman Imperial times, especially in Western Europe. They, too, are of fist size or a little bigger, but hollow with circular holes in the faces and knobs on the vertices (Figure 4). Nothing is known for certain about their use. A recently found specimen (Cervi-Brunier [7]) has the names of the twelve signs of the zodiac inscribed on its faces. A photograph of another typical example is reproduced by Malkevitch [17, p.84].

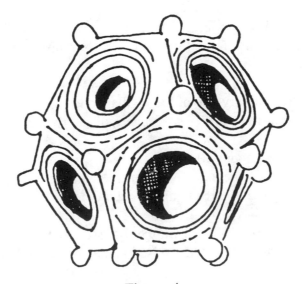

Figure 4

How could Hippasus have known about dodecahedra? Certainly not from Scottish stone balls or from Imperial Roman bronze pieces, but most likely from pyrite crystals. Pyrite was mined as iron ore, but more important was a common utensil used as a fire-starter in the bronze age and even earlier. Almost regular dodecahedral crystals of pyrite are abundant. Hence one might say that there is 'fire inside the dodecahedron'. This gives much credibility to a thesis by F. H. Thompson [24] who suggests that candles may have burnt inside the Roman dodecahedra. Aristotle, moreover, reports in *Metaphysics* A.3 (984a7) that Hippasus regarded fire as the first principle. This may be another reference, however vague, to the connection between fire and the dodecahedron. On the other hand, we have the signs of the zodiac inscribed on the specimen from Geneva. We will see in the next section that the association of the dodecahedra with astronomy or astrology would be more in accordance with Plato's views.

B. Concept and Classification: Theaetetus and the Five Solids, 380 B.C.

The anonymous scholiast to Euclid tells us about Theaetetus' discovery of the octahedron and the icosahedron. In accordance with a very detailed study by Eva Sachs [22] and a recent one by William Waterhouse [27], it is generally accepted that Theaetetus (1) - defined the concept of regularity, (2) - constructed the octahedron and the icosahedron, and (3) - proved that there are only five regular solids. Theaetetus created the prototype of a mathematical theory; starting from a significant example, the dodecahedron, he proceeds to a general concept, and manages to classify all objects satisfying the definition. A detailed study of the objects, surviving in the second half of Euclid's Book XIII, completes the investigation.

As pointed out by Waterhouse, Theaetetus' definition was a genuine act of mathematical creativity. There was nothing to generalize or extend - "The crucial discovery was the very concept of a regular solid" [27, p. 214]. What, exactly, was Theaetetus' concept? Most likely it was the first of the following three definitions surviving from antiquity.

1. Plato, *Timaeus* 55a (written about 370 B.C.?):
    "...a solid that gives rise to a subdivision of the [circum-] sphere into equal and similar [= congruent] parts."

In fact, Euclid constructs every one of the regular solids together with its circumsphere. The old name of the dodecahedron, moreover, was "sphere of the twelve pentagons."

2. Euclid XIII, 18a (written about 300 B.C.):
    "...no other figure, besides the said five figures, can be constructed which is contained by equilateral and equiangular figures equal to each other."

In his proof of this theorem, Euclid implicitly assumes the congruence of the vertices in addition to the congruence of the faces. Otherwise his theorem would be false. (For example, see Lichtenberg [15].)

3. Proclus, *Commentary to Euclid* [20, p. 158] (written about 450 A.D.)
    "For is it not the task of the geometer to inquire why it is that an indefinite number of equilateral polygonal figures can be inscribed in a circle, whereas in a sphere it is not possible to inscribe an indefinite number of polyhedra with equal sides and angles and composed of similar faces?"

In fact, in this passage Proclus quotes Geminus, who wrote in the first century B.C., so Proclus' date of 450 A.D. is misleading. Proclus (or Geminus) notices the analogy between dimensions 2 and 3. Moreover, in the paragraph preceding the quotation, he discusses an example of 'concept and classification'.

The five solids in Plato's philosophy

The five regular polyhedra are called 'Platonic solids' because they are mentioned in Plato's *Timaeus* 53-57. He gives the definition quoted above and associates the solids with the elements; the tetrahedron with fire, the octahedron with air, the icosahedron with water, the cube with earth, and the dodecahedron with 'the whole.' In the *Phaedo* 110b Plato associates the dodecahedron with the true picture of the earth as seen from 'above.' (Not to be confused with the 'element earth.')

Given Plato's well-known interest in geometry, the five solids may have been an outstanding example for his doctrine of 'idea and participation.' (For instance *Phaedo* 100bc.) The number 3 'participates' in 'the odd' (*Phaedo* 103e - 105a), and likewise each of the five solids 'participates' in the 'idea' of a regular solid. Moreover, the definition of the concept of regularity may have been a typical example of ascending from the particular to the more general and satisfactory concept as demanded in the *Phaedo* 101d, and, in a more philosophical sense, in the *Republic* 511ab. In Plato's *Theaetetus* 147b, Socrates explicitly praises the student Theaetetus for creating a general concept.

There are countless books and papers on Plato's philosophy, but Sachs [22] and Waterhouse [27] are essential. The importance of geometry for Plato's doctrine of ideas is emphasized by Mittelstrass [18]. For a more cautious viewpoint, see Burnyeat [6].

C.   Fanciful Enthusiasm
Fra Luca Pacioli, 1509 A.D.

The famous Renaissance painter, Piero della Francesca (1415-1492), wrote a treatise about 'the five solids'. His student, Luca Pacioli (1445-1515), translated, edited, and enlarged this book and published it in 1509 under the title *Divina Proportione* [19], claiming that this was "...a work essential to all open spirits curious for knowledge." [1] The pictures in this book were provided by his friend, Leonardo da Vinci.

Pacioli published some more mathematical books "...from compassion with the ignorant." All of them are rather elementary. But the point is that he and the other mathematical authors of his generation revived Greek mathematics and made it accessible to their students, the famous Italian Renaissance algebraists. The mathematics in the *Divina Proportione* consists essentially of excerpts from Euclid. Pacioli stresses the mathematical importance of the golden section (divina proportione, in his words), but strictly confined to mathematics. Here are some of his chapter headings.

---

[1] I wish to thank Peter Hilton for the translation of this and all other quotations from Pacioli.

XXII.   "Of its thirteenth and most important elaboration. How without the knowledge of
         this proportion the construction of the regular pentagon is impossible. How Euclid
         in his proofs applies only that which precedes and not that which follows."

XXIV.   "How the stated elaborations contribute to the classification of all regular bodies and
         those dependent on them. Why these five bodies are called regular."

XXV.   "That it is impossible for there to be more than five regular bodies and why. That
         it is impossible to construct regular solids from hexagons, heptagons, octagons,
         nonagons, decagons, and other similar polygons."

Because mathematical instruction was based mainly on Euclid from the late medieval times
onward (as it had been in antiquity), knowledge of the five regular solids was widespread. Johannes
Kepler used them for the description and mathematical foundation of his cosmographic system, and
Descartes and Euler studied their combinatorial properties. Details about the work of these authors
can be found in the articles by Malkevitch [17] and Toepell [25].

Symmetry and the Golden Section

In modern mathematics, symmetries are defined as structure-preserving mappings of a
mathematical object onto itself, or, briefly, automorphisms. We will study such mappings of the
regular polyhedra in the next section.

When the Greeks called two segments symmetric, they meant that they 'had a common
measure' (see the definition in Euclid's Book X). That is, two segments A and B are called
commensurable (= symmetric), if there exists a segment D and natural numbers k and n such that
$A = kD$ and $B = nD$. As an easy consequence, Euclid proves in Book X, 5 and 6, that two segments
A and B are commensurable if and only if they have to one another a ratio which a number has to
a number, that is, if there exist natural numbers k and n such that $A/B = k/n$. Using this criterion,
Pacioli says the following about the golden section.

XV.   "Of its sixth unnameable elaboration. How no rational number can be so divided in
         this proportion that the parts are themselves rational."

Pacioli credits Campanus with the following proof, which I present in modern terms. Let d and f
be the diagonal and side of a regular pentagon, as in our section A, such that $d/f = f/(d-f)$. If d and
f were commensurable, then we could write $d/f = k/n$ and take k/n in lowest terms (relatively prime)
by Euclid VII, 20, 21, 22. Now $d/f = k/n = f/(d-f) = n/(k-n)$ in lower terms, a contradiction. (One
could also use Euclid IX, 16.) Hence, in Greek terms, the golden ratio is completely
'unsymmetric.'

The Greeks called the golden ratio the 'division in extreme and mean ratio' (Euclid VI,
Def. 3). Pacioli speaks of the 'divine proportion'. The term 'golden ratio' (= goldner Schnitt)
seems to have originated in Germany in the first half of the 19th century. The German art historian
A. Zeising was the first to 'discover' the importance of the golden section for art history in his

book *Aesthetik* [28]. As far as I can see, his claims of almost magic effects of the golden section have never been taken seriously by his fellow art historians. Nevertheless, they are popular with many people who delight in a mathematical explanation of beauty.

D. Abstraction:
Felix Klein and the Symmetry Groups of the Polyhedra, 1880

Roughly between 1800 and 1870, crystallographers and mathematicians introduced the concept of a group into mathematics. Felix Klein, who was equally interested in group theory and complex function theory, wrote his famous treatise, *The Icosahedron and the Solution of Equations of the Fifth Degree*, in 1884 [13]. At the very beginning of this book he studies the symmetry groups of the regular solids in detail. These are finite subgroups of the special orthogonal group SO_3 of rotations of 3-space. Klein gives credit to Laguerre and Cayley for establishing the isomorphism of SO_3 with the group PSU_2 of homogeneous, linear, unitary transformations of determinant 1 of the extended complex plane. This isomorphism connects geometry and function theory, thus giving rise to a new and very deep extension of the theory of regular polyhedra which flourishes to this day. (See [4] and [14].)

Starting from geometrical figures, Klein finds the following finite subgroups (up to conjugacy and together with their respective subgroups) of SO_3:

- dihedral groups as the symmetry groups of dihedrons or double-pyramids,
- the tetrahedral group, which is isomorphic to the group A_4 of even permutations of four symbols,
- the octahedral group, which is the same as the group of the cube, isomorphic to the full group S_4 of permutations of four symbols,
- the icosahedral group (= dodecahedral group), which is isomorphic to the group A_5 of even permutations of 5 symbols.

For the isomorphism between the dodecahedral group and A_5, Klein uses as permuted symbols the five cubes which can be inscribed into a regular dodecahedron (as already observed by Hippasus and Euclid). By geometric arguments Klein proves the simplicity of the dodecahedral group. This proof is reproduced with minor modifications by Artmann [2].

In an earlier paper of 1875, Klein states the completeness of the list of finite subgroups of SO_3 given above as a theorem. More generally, he considers finite subgroups of the groups of all isometries of euclidean 3-space. He proves that such a group must be a subgroup of SO_3 and concludes "...und also sind durch die oben angeführten Gruppen alle in Betracht kommenden Gruppen erschöpft" ("...and hence there are no other examples except the ones mentioned above"). He furnishes the proof for the subgroups of SO_3 subsequently in Chapter V, Section 2, of the book on the icosahedron [13, pp. 128-130 of the English edition]. Nice elementary presentations of this proof are given by Coxeter [9, pp. 70-72] and by Senechal [23].

Klein's theorem puts the theory of regular solids in a totally new perspective; the abstract groups of symmetries suffice to recapture the polyhedra. The dihedral subgroups of SO_3 have an invariant subspace (the axis of the double pryamid), and all other finite subgroups of SO₃ are (conjugate to) subgroups of the symmetries of the tetrahedron, the octahedron, or the icosahedron.

The Dodecahedron again

The elementary building blocks of finite groups are the simple groups. These are groups without proper homomorphic images, or, equivalently, without proper normal subgroups. Groups of prime order p are simple, and these are the only abelian simple groups. The smallest non-abelian simple group has order 60; it is the symmetry group of the dodecahedron (or icosahedron). As mentioned above, this group is isomorphic to the alternating group A_5, and thus is the first one in the series of simple groups A_n ($n \geq 5$). Several other series of finite simple groups are known, and, moreover, there are some so-called "sporadic" simple groups. In one of the most exciting stories in modern mathematics, in 1960 - 1980 group theorists completed the classification of all finite simple groups. (For details see [8].) Like at the beginning of Greek mathematics, the dodecahedron has again provided the first significant example for a great mathematical theory.

References

[1]   Artmann, B., "Hippasos und das Dodekaeder," *Mitt. Math Sem. Gießen*, 163, (Coxeter-Festschrift, 1984), pp.103-121.

[2]   Artmann, B., "A simple proof for the simplicity of A_5," *The American Mathematical Monthly*, Vol 95, n.4 (1988), pp. 344-349.

[3]   Artmann, B., "Euclid's ELEMENTS and its pre-history," *Apeiron* XXIV, no.4 (1992), pp. 1-48.

[4]   Bättig, D. and H. Knörrer, "Singularitäten," Basel: Birkhauser (1991).

[5]   Burkert, W., "Lore and Science in Ancient Pythagoreism," Cambridge, Mass. (1972).

[6]   Burnyeat, M.F., "Platonism and Mathematics, A Prelude to Discussion," In A. Graeser, ed., "Mathematics and Metaphysics in Aristotle," Bern: Paul Haupt (1987), pp. 213-240.

[7]   Cervi-Brunier, I., "Le dodécaèdre en argent trouvé à Saint Pierre de Genève," *Zeitschrift für Schweizerische Archäologie und Kunstgeschichte*, (Band 42, 1985, Heft 3), pp. 153-156.

[8]   Conway, J.H., "Monsters and Moonshine," *The Mathematical Intelligencer 2*, no.4 (1980), pp. 165-171.

[9]   Coxeter, H.S.M., "Regular and Semi-regular Polyhedra," In M. Senechal & G. Fleck, eds., "Shaping Space, A Polyhedral Approach," Birkhäuser: Basel (1988).

[10]   Critchlow, K., "Time Stands Still," London: Gordon Fraser (1979).

[11]   Euclid-Heath, "The Thirteen books of Euclid's Elements," 3 vols, New York: Dover (1956).

[12]   von Fritz, K., "The Discovery of Incommensurability by Hippasos of Metapontum," *Annals of Mathematics 46* (1945), pp. 242-264.

[13]   Klein, F., "Vorlesungen über das Ikosaeder und die Auflösung der Gleichungen vom fünften Grade," Leipzig: Teubner (1884), English translation, New York: Dover (1956).

[14]   Lamotke, K., "Regular Solids and Isolated Singularities," Wiesbaden: Vieweg (1986).

[15]   Lichtenberg, D.R., "Pyramids, Prisms, Antiprisms, and Deltahedra," *Mathematics Teacher* (April, 1988), pp. 261-265.

[16]   Lindemann, F., "Zur Geschichte der Polyeder und der Zahlzeichen," *Sitzungsber, der math.-phys. Klasse der Kgl. baierischen Akad. d. Wiss. XXVI* (1896), pp. 625-783.

[17]   Malkevich, J., "Milestones in the history of polyhedra," In M. Senechat & G. Fleck, eds., "Shaping Space, A Polyhedral Approach," Birkhäuser: Basel (1988).

[18]   Mittelstrass, J., "Die geometrischen Wurzeln der platonischen Ideenlehre," *Gymnasium 92* (1985), pp. 399-418.

[19]   Pacioli, Fra Luca, "Divina Proportione," Venedig (1509), New herausgegeben, übersetzt und erläutert von Constantin Winterberg, Wien (1889).

[20]   Proclus (Morrow, G.R.), "A Commentary on the First Book of Euclid's Elements," Translated by G.R. Morrow, Princeton: Princeton Univ. Press (1970).

[21]   Ritchie, Graham and Anna, Scotland (Archeology), London (1981).

[22]   Sachs, E., "Die fünf platonischen Körper," Berlin (1917).

[23]   Senechal, M., "Finding the finite groups of symmetries of the sphere," *Amer. Math. Monthly*, vol. 97 (April, 1990), pp. 329-335.

[24]   Thompson, F.H., "Dodecahedrons Again," *The Antiquaries Journal*, vol. 2, part I (1970), pp. 93-96.

[25]   Toepell, M., "Platonische Körper in Antike und Neuzeit," *Der Mathematikunterricht 37*, Heft 4 (1991), pp. 45-79.

[26]   van der Waerden, B.L., "Die Pythagoreer," Zürich: Artemis (1979).

[27]   Waterhouse, W.C., "The discovery of the regular solids," *Arch. Hist. Exact Sciences 9* (1972-1973), pp. 212-221.

[28]   Zeising, A., *Aesthetik*, Leipzig (1855).

G. A. Miller: Mathematician, Historian of Mathematics, and Mathematics Educator Extraordinaire

Richard M. Davitt
University of Louisville
Louisville, KY 40292

George Abram Miller's tenure in the mathematics department at the University of Illinois spanned 44 years from his arrival on campus as an already established researcher in finite group theory in 1906, through his retirement from teaching at age 68 in 1931, until his death in February, 1951, from the complications of old age and sixteen months of bereavement over the loss of his beloved wife, Cassandra, his helpmeet and companion for all but four of those fruitful years. At Illinois students and his colleagues had long known him to be a modest, unassuming, charming human being. Over the years he was more and more perceived to be a quiet eccentric with an international mathematical reputation who even in retirement came seven days a week to work in his notoriously cluttered cubbyhole of an office on the third floor of the Mathematics Building. In actual fact, although a professor emeritus, he still did group theoretic research, helped with the editing of his *Collected Works*, and continued to write journal articles dealing with mathematics education, the popularization of mathematics, and his longstanding second field of interest, the history of mathematics. For nearly 20 years the shabby, white-haired, diminutive "Papa" Miller had lived frugally in retirement on a pension of about $250 a month. His wardrobe consisted largely of two rumpled suits which he had been wearing, summer and winter, for as long as anybody could remember. His overshoes - flopping strips of rubber precariously tied together with pieces of string and rubber bands - were a campus legend. He enjoyed the company of young students and usually took his lunch and sometimes his dinner in the student cafeteria of the Illini Union close by his office. His lone concession to technological progress was the purchase of an automobile in 1921. He had it for a dozen years but never had much use for it nor enjoyed driving it. When a state law was enacted requiring cars to have rear view mirrors, Miller observed with quiet satisfaction that his had none and promptly sold it for $75 with less than 10,000 miles registering on its odometer. In any case he loved to walk, both the short distance between his apartment and his office and on long, meditative forays around the Illinois campus. Lost in his mathematical dreams, he always walked with his hands firmly clasped behind his back, oblivious to everyone and everything around him.

The Millers had no children and their simple social life consisted primarily in entertaining and being entertained by a close circle of friends in home settings. Although they occasionally visited Mrs. Miller's relatives in Danville, they took only two extended vacations in forty years, one of which was their honeymoon. Miller stated simply that he disliked vacations "because he couldn't get any work done."

Throughout their married life they lived in a six room apartment half a block from campus except for a two year period around the time of Miller's retirement when, at Mrs. Miller's insistence, they moved into a small bungalow six blocks from campus. Eventually they went back to their old apartment where service was provided because the professor was hopeless as a home owner. He could not cope with the furnace nor the lawn mower and had no patience with leaky faucets.

Previous to his retirement Miller had always been a popular, well-liked teacher who seldom flunked anybody. When a new student entered his course on group theory, Professor Miller would give him 38 cents and order him to buy a textbook on group theory coauthored by George A. Miller. In this way the Professor, who received a royalty of 38 cents on each text, avoided making money off of his students.

Despite being frugal, Miller was not miserly. Beyond satisfying simple wants for food, clothes, and shelter, he had no real use for money. He had long been known to be a soft touch for students who were in financial difficulties and to be quietly generous in his financial support of his church, university, and community. Unbeknownst to his colleagues, students, and many believe, even his wife, Miller had long been investing his savings with the same determined, patient approach that characterized his tedious, painstaking work in finite group theory. The quiet little man with a briefcase who had been visiting Miller regularly at his office for some 25 years turned out to be A. Burkey Gwinn, an investment counselor, who had helped the professor parlay his outwardly meager financial resources into almost $1,000,000 by the time of his death. Indeed Miller further shocked everyone except his lawyer and banker when he bequeathed this entire amount to the university because, as he told his lawyer when his will was prepared, "Everything I have I received from the university, and I simply want to repay my obligation."

How he did it was at first a mystery. His largest annual salary was $6000 and his average salary at Illinois was only $3300. He inherited no money and had no income except his university salary. However, he always had an unshakable faith in small business and early in his career he began to put his savings into the stocks of small companies; with his profits he bought more. Staying the course through the depression and the inflationary post-war years, he eventually was able to bequeath to Illinois the largest liquid asset it had ever received. His will specified only that the university was to use his gift for educational purposes, teaching and research, and that it could not be used for current general operating expenses.

Since his death the university has put his bequest to good use, using its income, for example, to initiate the *Illinois Journal of Mathematics* in 1957 and to subsidize its publication since then. This income also in large part supports the work of the university's Miller Endowment Committee which provides grants to all academic units for Visiting Professors and Scholars and sponsors an all-campus lecture series. Year after year these programs greatly enhance the intellectual and cultural life at Illinois making them an especially appropriate memorial to a man who believed so deeply in the value of a liberal education.

G. A. Miller's life story before he came to Illinois has the same "Horatio Alger" flavor as did his career there. He was born on a farm near Lynnville, Pennsylvania, on July 31, 1863, the sixth of seven children of thrifty, hardworking, self-sufficient Pennsylvania Dutch parents. His family spoke German at home but his schooling was conducted in English. There was no family tradition nor expectation that young George or his siblings become highly educated. Nevertheless, when he found himself unable to continue his education without self-support, Miller began to teach school at the age of seventeen. Eventually he obtained his Bachelor of Arts from Muhlenberg College in nearby Allentown, Pennsylvania, in 1887. During the first year after graduation he moved with his brother and parents to Greeley, Kansas, where he was a Principal in the public schools.

From 1888 to 1893 he was Professor of Mathematics at Eureka College (Illinois). Although his most advanced work in mathematics had been in calculus, in 1890 he began to offer advanced mathematics courses for post-graduates leading to the doctorate. Although he spent part of the summer of 1889 at Johns Hopkins and the summer of 1890 at Michigan, neither university was in session during those periods. There is no evidence that he learned anything about abstract groups from Bolza at Johns Hopkins nor F. N. Cole at Michigan at that time. Although he was steadily expanding his offerings at Eureka, a course in group theory was not among them. In 1891 he enrolled by correspondence as a graduate student in Cumberland University, Lebanon, Tennessee, and received his doctorate from there just one year later. It was then possible to do all one's course work by correspondence and to substitute examinations in advanced courses for thesis requirements. It is highly probable that Miller never visited Cumberland in light of the fact that he was teaching six hours a day at Eureka during that year.

Miller's first introduction to group theory and research level mathematics took place from 1893 to 1895 while he was an instructor of mathematics at Michigan. He learned the fundamentals of that subject from F. N. Cole, and even lived in his house for two years. He then spent the following two years in Europe where he studied group theory, first under S. Lie at Leipzig, and subsequently with C. Jordan in Paris. The specializations of neither man appealed to Miller, but he did adapt for himself Lie's ideas about commutators and commutator subgroups and Jordan's concerns about the primitivity/imprimitivity of finite substitution groups. Using his own methods and insights into researching the structures of finite groups, he had already published 27 papers in the mathematical journals of America, England, France, and Germany by the end of the year of his return.

From 1897 to 1901 he was an assistant professor at Cornell and from 1901 to 1906 he was first an assistant and then an associate professor at Stanford. By the time he was recruited to Illinois in 1906 by the chair of the mathematics department, E. J. Townsend, he already had over 100 articles in print and had received a prize from the Academy of Sciences of Cracow for his determination of the substitution groups of degree ten. The prize had been standing since 1886 and was the first such award to an American for work in pure mathematics.

By the end of his career Miller had published upwards of 820 papers in more than forty different educational, scientific, and mathematical journals of eleven countries. Approximately 450 of them made direct contributions to the theory of finite groups. These technical papers were all directed at the fundamental problem of such groups, namely, the determination of what groups exist and methods of distinguishing one group from another. He was always looking for a key that would unlock the door to groups of order n. In 1902, for example, he gave a method for the determination of all groups of order p^n, p a prime. In 1930 he used that method to list for the first time the 294 groups of order 64. He tried the attack of determining orders n for which a specified small number of groups exists. Thus a large number of his papers determine the groups which have rather strange sets of properties; these are motivated by the desire to find out what sorts of questions one needs to ask of a group in order to be able to identify it. Few mathematicians have attempted to extend his work using his methods, yet very little of it has been extended by other methods. Since his death, on the global level, groups of odd order have been shown to be solvable and the determination of all sporadic simple groups has been made, but on the local level a fairly long list of information is still needed before a specific finite group can be uniquely determined.

Many basic results in finite group theory were established by Miller. For example, he was the first to prove the simplicity of the multiply transitive Mathieu groups of degrees 22, 23, and 24 and to fully develop such concepts as the commutator subgroup, the derived series, the Frattini subgroup, and the characteristic subgroups. All of his group theory papers, a selection of his other publications, and a number of historical accounts detailing the history of finite group theory during the various chronological periods of Miller's research career appear in *The Collected Works of George Abram Miller* which were published in five volumes by the University of Illinois in 1935, 1938, 1946, 1955, and 1959. The first three volumes were published under the editorship of a three person committee consisting of H. R. Brahana, R. D. Carmichael, and A. B. Coble with Miller himself doing most of the actual work. H. R. Brahana was the sole editor who saw the project through to its completion after the death of Miller and the retirement of Carmichael and Coble. In addition to their substantive contents, the indices and appendices of these volumes offer a complete bibliography of Miller's writing. Contrary to what Garrett Birkhoff claims [14], these collected works were not published as a by-product of Miller's philanthropy but spontaneously by Illinois as a "fitting memorial of his contributions to mathematical scholarship and to the renown of the University" [10, Preface to Volume I] upon his retirement from the teaching faculty. Furthermore, Birkhoff's additional comment that "I think it fair to say that the publication was not in response to irresistible public demand." [14], is an uncalled for slap in the face of a man who for a period of a quarter of a century was regarded as the preeminent finite group theorist in the world and one of the very first American mathematicians of truly international reputation.

Miller published two books, *Determinants* (1897), written for the class in advanced algebra at Eureka, and *Historical Introduction to Mathematical Literature* (1916). He wrote part of each of two other books, "The algebraic equation," (pp. 211-260) of *Monographs in Topics of Modern Mathematics* (1911), edited by J. W. A. Young, and *Finite Groups* (pp. 1-192) (1916) by Miller, H. F. Blichfeldt, and L. E. Dickson.

Miller was ever a proponent of the value of professional societies and their sundry efforts to improve the lot of mathematics and science. In October, 1891, he was elected to membership in the New York Mathematical Society which had had a roster of just twenty three members at the beginning of that year. This Society, of course, became the American Mathematical Society in 1894. He was a member of its Council from 1901 to 1904. He was a founding member of the San

Francisco Section and its Secretary until 1906. He was Vice President of the Society in 1908 and Chairman of the Chicago Section from 1907 to 1909. His papers appear in the first volumes of both the *Bulletin* (1894) and the *Transactions* (1900) of the AMS.

His relationship with the *American Mathematical Monthly* and the Mathematical Association of America, the organization which eventually adopted it as its official publication is also quite close. In Volumes 2 and 3 of the *Monthly*, (1895, 1896) he published in a series of fourteen installments some forty pages of exposition of substitution groups. Thereafter he contributed seventy more papers to the *Monthly*, the last in 1933 when he was seventy years old. In 1909 the University of Illinois joined the University of Chicago in support of the *Monthly* and Miller was recruited to be one of its editors. He remained on the editorial board until 1915. At the April, 1914, meeting of the Chicago Section of the AMS, Miller was appointed to a five person committee whose charge was to make recommendations to the Council of the Society relating to the perceived need to improve the situation of collegiate mathematics. Subsequently, the Council set up its own committee to consider the same issue. Eventually the AMS decided not to enter into the activities of the special field covered by the *Monthly* but to recommend the formation of a new and separate organization to promote the cause of secondary and collegiate mathematics. Thus the MAA was founded in Columbus, Ohio, on December 30, 1915. Miller was one of the 104 people in attendance at that meeting and indeed was elected one of two Vice-Presidents of the MAA for 1916. He became its sixth president in 1921 and was an active participant in its national and regional meetings for many years.

Miller was also a vigorous supporter of the National Academy of Sciences to which he was elected a member in 1921. He published two papers in the first volume of the *Proceedings of the Academy* in 1915 and another two dozen or so before his retirement in 1931. Thereafter he contributed five or six papers to each volume of the *Proceedings* for many years; his last two mathematical papers appeared in Volume 32 in 1946.

Additionally Miller was a fellow of the American Academy of Arts and Sciences and an active member of the American Association for the Advancement of Science. In 1914 he, along with E. H. Moore, was appointed to be the representative for Pure Mathematics on the Committee of 100 on scientific research.

To quote Brahana [4],

Miller's most important publication on a general subject was probably "Some Thoughts on Modern Mathematical Research." This lecture was given before the University of Illinois chapter of Sigma Xi in 1912. It was printed in *Science* and reprinted in the *Annual Report of the Smithsonian Institution* as one of the most important memoirs published in America during the year. "Some may be tempted to say that the useful parts of mathematics are very elementary and have little contact with modern research. In answer we may observe that it is very questionable whether the ratio of the developed mathematics to that which is finding direct application to things which relate to material advantages is greater now than it was at the time of the ancient Greeks." These sentences from the concluding paragraph suggest the tone; they are arresting today.

A similar article addressed to the broader scientific community was entitled "The Function of Mathematics in Scientific Research." It was also published in *Science* (1917) and was a lecture given to the Science Club of the University of Wisconsin. In it he asks the question, "What is the attitude of mind which has contributed most powerfully to mathematical progress?" to which he answers, "An attitude of modesty, discretion, and a love for mental travel and exploration." He goes on to state that, vis-a-vis scientific research, mathematics must discover and develop unifying processes which are sufficiently comprehensive to avoid bewilderment as a result of many details, and yet sufficiently close to the concrete to become useful in widely separated fields of scientific endeavor. He strongly espouses the position that great latitude should always be permitted the mathematical researcher because of the essentially infinite number of avenues his investigations might take.

In his papers touching upon mathematics education he was a strong proponent of teachers of secondary mathematics using more advanced topics such as groups in their classrooms. In "A New Chapter on Trigonometry," for example, he demonstrated how the octic group could be used to generate new values of trigonometric functions from previously calculated values. He also espoused the integration of mathematics education majors into the university mathematics classroom, worked to promote summer institutes for secondary mathematics teachers, and suggested specific programs by which the professional mathematics community could continue to replenish itself. A typical article along these lines is entitled "External Encouragement for the Study of Higher Mathematics" and appeared in the *Monthly* in 1908. What Miller was trying to do has a very contemporary feel; the mathematical community is still trying today to bring the ideas of higher mathematics to the attention of secondary teachers and within their reach.

Miller's interest in the history of mathematics dates from his time at Eureka; thus he studied history longer than he studied groups. His first publication on a historical subject was in 1901, and his last in 1947. His four greatest personal contributions to the field were:

(1)   The set of articles he wrote on the history of finite groups for Volumes I, II, and III of his *Collected Works*. These give a detailed and fairly accurate account of such groups from the time they were first investigated up to 1915.

(2)   His *Historical Introduction* which was in reality a loosely organized, topical history of mathematics.

(3)   A series of eleven "Lessons" in the history of mathematics which he published in various issues of *Mathematics Magazine* between 1939 and 1947.

(4)   His unpublished *Introduction to the History of Elementary Mathematics*. When this work was offered for publication in the early 1930's there was no market for such a text. Although the editorial committee of his *Collected Works* and H. R. Brahana in particular had indicated that it would be published in the last volume of that series, this did not occur. However, Miller's working manuscript copy of this text can be found in the Miller file in the Archives of the University of Illinois.

Throughout his career Miller was a severe critic of historical methodology in mathematics and a determined zealot when it came to rooting out errors in mathematics and its history wherever he detected them. Perhaps the best summary of Miller's philosophical position on the role of a reviewer and also that of a student when considering historical tracts has been given by G. Waldo Dunnington, a colleague of Miller at Illinois and the long-time editor of the Humanism and History of Mathematics column of *Mathematics Magazine*. In his article "G. A. Miller as Mathematician and Man: Some Salient Facts" in that column [6] he wrote:

> As a reviewer Professor Miller has always taken the position that the truth should be told. He feels that misinformation is worse than none at all, he is thoroughgoing, uncompromising with error and intellectual dishonesty wherever found. He goes to great lengths either to verify or disprove statements, tireless in the search for reliable accurate information, yet unfailingly kind and appreciative - giving praise to real merit where this is due. When praise comes from so fearless and honest an authority one knows it is deserved and the recipient feels correspondingly encouraged. To him a review is, what it should be, an impartial discussion of (or contribution to) the subject.
>
> In treating the history of mathematics Professor Miller has stated a number of fundamental desiderata, at various times, which might well be summarized here: (1) There should be a clear definition of terms involved. (2) The reader should be encouraged to draw his own conclusion from the evidences presented and should not be confronted merely with a conclusion which he is expected to accept. (3) The tendency towards deviating from the ordinary language and the ordinary grammatical constructions in speaking about mathematical questions should be noted. (4) The reader must adopt the policy of not adopting statements without verification. (5) The beginner must differentiate between working hypotheses and established facts. (6) Many of the advances in the history of mathematics as well as mathematics itself must be due to the repeated correction of errors. (7) Historical statements are frequently rich in their implications so that the student should often draw a large number of conclusions from a single statement. Miller's *A Historical Introduction to Mathematical Literature* (1916) has become a classic.

It is of some interest to compare the style and format of each of Miller's efforts to survey broadly the history of mathematics. All three are organized loosely by chapters or lesson, with articles numbered consecutively by specific topic. There is surprisingly little overlap in each of these efforts. The table of contents for *Historical Introduction* and *History of Elementary Mathematics*, the contents of the eleven Lessons and several typical Chapter subdivisions are reproduced in the appendix to this paper. Perusing them it is easy to taste the flavor of what Miller offered his reader in the way of historical information.

Unfortunately Miller's besetting sin as a historian was that he did not generally feel called upon to acquaint himself at firsthand with original sources. He often scolded other historians for errors and inaccuracies he detected in their work yet did not truly practice everything he preached.

Having noted this let it be said that seldom in the history of mathematics has there been an individual more dedicated to doing mathematics, studying its history and spreading its grand message than G. A. Miller. All who knew him would agree that he lived and breathed mathematics throughout his prolific mathematical career. It is unlikely that the world will ever see his like again.

References:

(1)   Archibald, R. C.,   *A Semicentennial History of the American Mathematical Society, 1888-1938*, American Mathematical Society, New York, 1938.

(2)   Brahana, H. R.,   "George Abram Miller,"   *American Mathematical Monthly*,   58(1951), 447-449.

(3)   Brahana, H. R., "George Abram Miller 1863-1951," *Bulletin AMS*, 57(1951), 377-382.

(4)   Brahana, H. R., George Abram Miller, 1863-1951," *Biographical memoirs, National Academy of Sciences.* 39(1957) 256-312.

(5)   Cajori, Florian, *A History of Mathematics, Second Edition*, The Macmillan Company, London, 1919.

(6)   Dunnington, G. Waldo, "G. A. Miller as Mathematician and Man: Some Salient Facts", *Mathematics Magazine*, 12(1938), 384-387.

(7)   Eisele, Carolyn, "Miller, George Abram", *Dictionary of Scientific Biography*, Charles Scribner's Son, New York, 1981, Vol. 9, 387-388.

(8)   Fay, Bill, "Millers Million", *Collier's for July 21*, 1951, pp 15, 62, 64, 66.

(9)   May, Kenneth O., Editor, *The Mathematical Association of America: Its First Fifty Years*, The Mathematical Association of America, 1972.

(10)   Miller, G. A., *The Collected Works of George Abram Miller, 5 Vols.*, University of Illinois, Urbana, Illinois, 1935-1959.

(11)   Miller, G. A., *Historical Introduction to Mathematical Literature*, The Macmillan Company, New York, 1921.

(12)   Miller, G. A., *Introduction to the History of Elementary Mathematics*, Unpublished, Archives, University of Illinois, Urbana, Illinois, c. 1934.

(13)   Tarwater, Dalton, Editor, *The Bicentennial Tribute to American Mathematics, 1776-1976*, The Mathematical Association of America, 1977.

(14)   Tarwater, J. D., White, J. T., and Miller, J. D., Editors, *Men and Institutions in American Mathematics*, Graduate Studies, No 13, Texas Tech University, Lubbock, Texas.

Appendix:

Historical Introduction to Mathematical Literature
(Published in 1916)

A Lesson in the History of Mathematics
(Published irregularly in the *National Mathematics Magazine's*
"History of Mathematics" Column, Vol. 13-21, 1939-1947)

Lesson	Topics
First	1. One half; 2. Sexagesimal system; 3. Abstractions
Second	4. Differences in attainments; 5. Decimal system; 6. Generalizations
Third	7. Quadratic equations; 8. Approximations regarded as exact; 9. Fundamental laws of combination
Fourth	10. Greek constructions; 11. Homogeneity
Fifth	12. Cubic equations; 13. Complex numbers
Sixth	14. The circle and the sphere; 15. Regular polygons
Seventh	16. The triangle
Eighth	17. Real numbers
Ninth	18. Infinity
Tenth	19. Theory of groups
Eleventh	20. Questionable historical statements

History of Elementary Mathematics
(Unpublished - Written c. 1934)

Table of Contents:

The History of the Mathematical Representation of Electromagnetic Theory

Domina Eberle Spencer   Uma Y. Shama
University of Connecticut   Bridgewater State College
Storrs, CT 06269   Bridgewater, MA 02325

1.   Before 1820

Before 1820, three fields were well known: the gravitational, the electrostatic and the magnetostatic. In 1666, while an undergraduate at Cambridge, Newton had discovered that the gravitational force between two masses could be described by the equation

$$\vec{F}_2 = -\,G\,\frac{m_1 m_2}{r^2}\,\vec{a}_r\,. \tag{1}$$

However, Newton did not publish [1] his results until 1685. Nearly a century later in 1777, Lagrange [2] discovered that the gravitational field could be expressed in terms of a scalar potential. In modern vector notation Lagrange's result is written

$$\vec{\mathcal{F}} = -\,\operatorname{grad}\phi \tag{2}$$

where $\vec{\mathcal{F}}$ is the force per unit mass in the gravitational field (the acceleration of gravity). In 1785, Laplace [3] discovered that this scalar potential satisfies the differential equation $\nabla^2\phi = 0$ in a region in which there is no matter. Exactly a century after the publication of Newton's equation, Coulomb [4] suggested that the electrostatic field could be described by the equation

$$\vec{F}_2 = \frac{1}{4\pi\epsilon_0}\frac{Q_1 Q_2}{r^2}\vec{a}_r\,. \tag{3}$$

Here the Coulomb equation is written in vector notation in the mks system. In 1813 Poisson [5] showed that in a region in which the density of matter is δ, the differential equation satisfied by the gravitational potential is $\nabla^2\phi = 4\pi G\delta$. In 1824, Poisson [6] showed that Laplace's equation also described the magnetostatic field in a current free region.

Thus, in 1820, all of the known physical fields could be described in the same mathematical form by the equation

$$\vec{\mathfrak{F}} = \vec{a}_r \frac{\mathrm{K}}{r^2}.$$
(4)

The line integral about a closed path was believed to always be zero.

$$\oint \vec{\mathfrak{F}} \cdot \vec{\mathrm{ds}} = 0$$
(5)

and all of the fields could be expressed as the gradient of a scalar potential. In modern vector notation we would say that curl $\vec{\mathfrak{F}} = 0$ and div $\vec{\mathfrak{F}} = -\nabla^2\phi$. If the divergence of the field vector vanishes, the scalar potential satisfies Laplace's equation. This occurs in the gravitational field in mass free regions, in the electrostatic field in charge free regions and in the magnetostatic field wherever the current density is equal to zero.

2. In 1820

In 1820, the neat world in which all known physical fields could be described in terms of a scalar potential was shattered by the work of Oersted and Ampère. Oersted [7] discovered that there was an interaction between the electric and magnetic fields in 1820. That same year Ampère discovered the interaction between two current carrying conductors. By 1823 Ampère[8] had done extensive experiments and had derived an equation for the force $\mathbf{d^2\vec{F}_A}$ between two infinitesimal current carrying conductors, $\mathbf{\vec{ds}_1}$ and $\mathbf{\vec{ds}_2}$ (See Figure 1).

$$\mathrm{d}^2\vec{\mathrm{F}}_\mathrm{A} = -\frac{I_1 I_2}{4\pi\epsilon_0 c^2 r^2}\vec{a}_r\left[2\left(\vec{\mathrm{ds}}_1 \cdot \vec{\mathrm{ds}}_2\right) - 3\left(\vec{\mathrm{ds}}_1 \cdot \vec{a}_r\right)\left(\vec{\mathrm{ds}}_2 \cdot \vec{a}_r\right)\right]$$
(6)

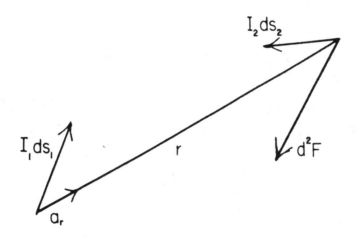

Figure 1
The force between current elements.

According to Maxwell [9], "The experimental investigation by which Ampère established the laws of the mechanical action between electric currents is one of the most brilliant achievements in science. The whole, theory and experiment, seems as if it had leaped full-grown and full-armed, from the brain of the Newton of electricity. It is perfect in form and unassailable in accuracy, and it is summed up in a formula from which all the phenomena may be deduced, and which must always remain the cardinal formula of electrodynamics."

3.   Gauss and Weber

Gauss was the first to attempt to delve deeper. In a letter [10] written in July, 1835, he said that the fundamental equation of electrodynamics should be an equation for the force between moving charges. Suppose that a charge Q (See Figure 2) moves at velocity $\vec{v}$. The equation for the force per unit charge $\vec{\mathcal{F}}_G$ on a test charge moving at velocity $\vec{u}$ suggested by Gauss was (if expressed in modern vector notation in the mks system of units)

$$\vec{\mathcal{F}}_G = \vec{a}_r \frac{Q}{4\pi\epsilon_0 r^2}\left[1 + \frac{(\vec{v}-\vec{u})\cdot(\vec{v}-\vec{u})}{c^2} - \frac{3}{2c^2}\left\{(\vec{v}-\vec{u})\cdot\vec{a}_r\right\}^2\right]. \tag{7}$$

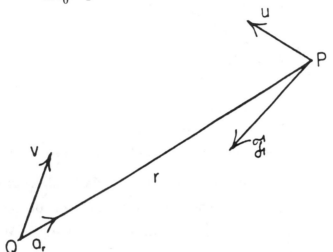

Figure 2
Force $\vec{\mathcal{F}}$ on a test charge at point P.

Like the fields known before 1820, this force could be expressed as the gradient of a scalar potential,

$$\phi_G = \frac{Q}{4\pi\epsilon_0 r}\left[1 + \frac{(\vec{v}-\vec{u})\cdot(\vec{v}-\vec{u})}{c^2} - \frac{3}{2c^2}\left\{(\vec{v}-\vec{u})\cdot\vec{a}_r\right\}^2\right]. \tag{8}$$

The Ampère expression for the force between current elements, Equation (6), can be derived from Gauss' equation for the force between moving charges. Gauss himself was sufficiently dissatisfied with this equation, that he never published it.

However, Gauss continued to write about the electrodynamic equation to his old friend, Weber, with whom he had invented the electric telegraph in 1833. In 1845, Gauss [11] suggested that the electrodynamic force equation should somehow include retardation. In 1848 Weber [12] presented the first published equation for the force between moving charges. Weber wrote the equation in the form

$$\vec{\mathcal{F}}_W = \vec{a}_r \frac{Q}{4\pi\epsilon_0 r^2}\left[1 - \frac{1}{2c^2}\left[\frac{dr}{dt}\right]^2 + \frac{r}{c^2}\frac{d^2r}{dt^2}\right] \tag{9}$$

which is equivalent to

$$\vec{\mathcal{F}}_W = \vec{a}_r \frac{Q}{4\pi\epsilon_0 r^2}\left[1 + \frac{(\vec{v}-\vec{u})\cdot(\vec{v}-\vec{u})}{c^2} - \frac{3}{2c^2}\{(\vec{v}-\vec{u})\cdot\vec{a}_r\}^2\right] - \vec{a}_r\frac{Q}{4\pi\epsilon_0 c^2 r}\left[\frac{d\vec{v}}{dt} - \frac{d\vec{u}}{dt}\right]\cdot\vec{a}_r . \tag{9a}$$

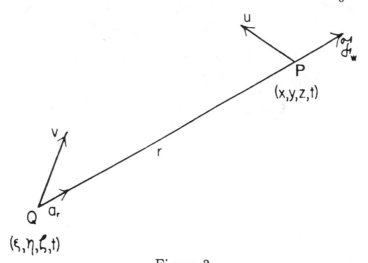

Figure 3

The Weber force $\vec{\mathcal{F}}_W$ on the test charge at point P produced by charge Q.

Note that for constant velocities Weber's equation is identical with the Gauss unpublished equation of 1835, Equation (7), yet Weber makes no mention of Gauss in his paper. This equation is still considered important by such modern scientists as Wesley [13], Pappas [14], Phipps [15], Graneau [16], and Assis [17].

4.   Faraday, Neumann, and Maxwell

The experimental basis for the next developments in electromagnetic theory was the work of Faraday. In 1821, he repeated all previous experiments [18]. In 1831 came his important paper on induction. Faraday's last paper [19] was published in 1855.

To describe the interaction of electric and magnetic fields demonstrated by Faraday's experiments, Franz Neumann [20] suggested in 1845 that the magnetic flux density $\vec{B}$ should be defined as the curl of a vector potential and that the electric field should be defined in terms of both a scalar potential ϕ and a vector potential $\vec{A}$.

$$\vec{B} = \text{curl }\vec{A}$$
$$\vec{E} = -\text{grad }\phi - \frac{\partial\vec{A}}{\partial t} \tag{10}$$

The same expression for $\vec{B}$ was developed independently in 1847 by Thomson [21]. The recognition of the necessity for the introduction of the vector potential is a landmark in the development of the mathematical representation of electromagnetic theory.

In the year in which Faraday published his last paper, Maxwell published his first paper [22]. The major aim of his scientific research was to develop the mathematical representation of Faraday's experiments. He wrote, "Faraday had the nucleus of everything electric since 1830." In 1861 Maxwell had developed his first equation

$$\operatorname{curl} \vec{E} = -\frac{\partial \vec{B}}{\partial t}$$

and had adopted the definitions of Neumann for $\vec{B}$ and $\vec{E}$. By 1864, Maxwell had developed the equation of conservation of charge,

$$\operatorname{div} \vec{J} = -\frac{\partial \rho}{\partial t}$$

and the expressions for the divergence of $\vec{D}$ and for the curl of $\vec{H}$

$$\operatorname{div} \vec{D} = \rho, \text{ and } \operatorname{curl} \vec{H} = \vec{J} + \frac{\partial \vec{D}}{\partial t}.$$

By 1868, he had decided that the scalar and vector potentials should be rejected and that the foundation of electromagnetic theory was the four equations

$$
\begin{aligned}
\operatorname{div} \vec{B} &= 0 & \operatorname{curl} \vec{E} &= -\frac{\partial \vec{B}}{\partial t} \\
\operatorname{div} \vec{D} &= \rho & \operatorname{curl} \vec{H} &= \vec{J} + \frac{\partial \vec{D}}{\partial t}
\end{aligned}
\tag{11}
$$

which we now call Maxwell equations. In 1873 Maxwell [23] published his *Treatise on Electricity and Magnetism* which is still regarded as the cornerstone of classical physics.

5. Lorenz, Heaviside, Lorentz, and Hertz

The potentials of Neumann were not retarded potentials. The Gaussian idea that retardation must somehow enter the theory was finally realized by Ludwig Lorenz [24] of Copenhagen who in 1867 introduced the retarded scalar and vector potentials for a continuous charge distribution

$$
\begin{aligned}
\phi(t) &= \frac{1}{4\pi\epsilon_0} \int \frac{\rho(\tau)}{r} \, d\mathcal{V} \\
\vec{A}(t) &= \frac{\mu_0}{4\pi} \int \frac{\vec{J(\tau)}}{r} \, d\mathcal{V}
\end{aligned}
\tag{12}
$$

where $\tau = t - r/c$. Lorenz showed that if the electric and magnetic fields were defined in terms of retarded scalar and vector potentials employing the Neumann definitions, Equation (10), then the Maxwell equations could be derived.

It was not until 1889 that Heaviside [25] suggested a link between the force per unit charge $\vec{\mathcal{F}}$ and the electric and magnetic fields. Heaviside proposed the equation

$$\vec{\mathcal{F}} = \vec{E} + \vec{u} \times \vec{B} \tag{13}$$

where $\vec{u}$ is the velocity of the test charge. This equation was also employed by Hendrik

Antoon Lorentz [26] of Leiden in 1892, who employed retarded potentials to describe the field of a moving electron:

$$\phi = \frac{Q}{4\pi\epsilon_0 r} \text{ and } \vec{A}(t) = \frac{\mu_0}{4\pi}\frac{Q\vec{v}(\tau)}{r} \tag{14}$$

where $\tau = t - \frac{r}{c}$.

Heinrich Hertz [27] suggested a small but very significant change in the definition of the electric field in terms of the scalar and vector potentials. Hertz proposed that the definition of Neumann, Equation (10), be modified to

$$\vec{B} = \text{curl } \vec{A}$$
$$\vec{E} = -\text{grad }\phi - \frac{d\vec{A}}{dt}. \tag{15}$$

Additional work on the development of the mathematical representation of electromagnetic theory was done in 1926 by Vanevar Bush [28], in 1954 by Moon and Spencer [29], in 1986 by Wesley [13] and by Phipps [15], in 1986 and in 1988 by Mirchandaney, Spencer, Uma and Mann [30].

6.   The Postulational Formulation of Electromagnetic Theory

The current state of the mathematical representation of electromagnetic theory can best be understood by studying its postulational development. The definitions of Neumann

$$\vec{B} = \text{curl } \vec{A} \text{ and } \vec{E} = -\text{grad }\phi - \frac{\partial\vec{A}}{\partial t} \tag{10}$$

or the definitions of Hertz

$$\vec{B} = \text{curl } \vec{A} \text{ and } \vec{E} = -\text{grad }\phi - \frac{d\vec{A}}{dt} \tag{15}$$

will be employed. The definition of Heaviside will be generalized to

$$\vec{\mathcal{F}} = \vec{E} + \vec{w} \times \vec{B} \tag{16}$$

where $\vec{w}$ is a velocity (either absolute or relative).

One of the Maxwell equations

$$\text{div }\vec{B} \equiv 0 \tag{17}$$

is an identity which follows directly from both the Neumann and Hertz definitions of the magnetic field in terms of a vector potential $\vec{A}$. The formulation of electromagnetic theory depends on four postulates; the postulated equations for ϕ, $\vec{A}$, $\vec{w}$ and the postulate on the velocity of light.

The postulates for the scalar potential of a point charge Q are either the Lorentz postulate

$$\phi(t) = \frac{Q}{4\pi\epsilon_0 (r)_R} \tag{18a}$$

or the postulate suggested by Liènard-Wiechert [31]

$$\phi(t) = \frac{Q}{4\pi\epsilon_0 (r)_R} \frac{1}{\left\{1 - \dfrac{\vec{v}(\tau) \cdot \vec{a}_{r_R}}{c}\right\}}. \tag{18b}$$

The vector potential for a moving point charge is defined by Lorentz as

$$\vec{A}(t) = \frac{Q\vec{v}(\tau)}{4\pi\epsilon_0 c^2 (r)_R} \tag{19a}$$

where $\vec{v}(\tau)$ is the retarded velocity of the source charge which produces the vector potential. An alternative form has been suggested by Liènard-Wiechert [31]

$$\vec{A}(t) = \frac{Q\,\vec{v}(\tau)}{4\pi\epsilon_0 c^2 (r)_R \left\{1 - \dfrac{\vec{v}(\tau) \cdot \vec{a}_{r_R}}{c}\right\}}. \tag{19b}$$

The Gaussian suggestion that only relative velocities should appear in electrodynamic equations suggests three other postulates;

$$\vec{A}(t) = \frac{Q\left[\vec{v}(\tau) - \vec{u}(t)\right]}{8\pi\epsilon_0 c^2 (r)_R}, \tag{19c}$$

where $\vec{v}$ is the velocity of the source at time $\tau = t - \frac{r}{c}$ and $\vec{u}$ is the velocity of the receiver at time t. Another possibility is

$$\vec{A}(t) = \frac{Q\left[\vec{v}(t) - \vec{u}(t)\right]}{8\pi\epsilon_0 c^2 (r)_R}, \tag{19d}$$

where both velocity of source and receiver are evaluated at time t. The last postulate for the vector potential is

$$\vec{A}(t) = \frac{Q\left[\vec{v}(\tau) - \vec{u}(\tau)\right]}{8\pi\epsilon_0 c^2 (r)_R}, \tag{19e}$$

where both velocity of source and receiver are retarded.

The classical postulate on the velocity term in the force equation, Equation 15, is

$$\vec{w} = \vec{u}(t) \tag{20a}$$

where $\vec{u}$ is the velocity of the receiver at time t. Gaussian formulations should employ relative velocities. Again there are three possibilities;

$$\vec{w} = \vec{u}(t) - \vec{v}(\tau) \tag{20b}$$

or

$$\vec{w} = \vec{u}(t) - \vec{v}(t) \tag{20c}$$

or

$$\vec{w} = \vec{u}(\tau) - \vec{v}(\tau). \tag{20d}$$

Three postulates on the velocity of light should be considered. For all of these

$$(r)_R = c(t - \tau). \tag{21}$$

The different postulates on the velocity of light employ different distances $(r)_R$ as illustrated in Figure 4.

Postulate I*: Einstein [32] - 1907

$$(r)_I = \left[[x(t) - \xi(\tau)]^2 + [y(t) - \eta(\tau)]^2 + [z(t) - \zeta(\tau)]^2 \right]^{1/2} \tag{22a}$$

or

Postulate II: Ritz [33] - 1908

$$(r)_{II} = \left[\left[x(t) - \xi(\tau) - \frac{d\xi(\tau)}{d\tau}(t - \tau) \right]^2 \right.$$
$$+ \left[y(t) - \eta(\tau) - \frac{d\eta(\tau)}{d\tau}(t - \tau) \right]^2$$
$$\left. + \left[z(t) - \zeta(\tau) - \frac{d\zeta(\tau)}{d\tau}(t - \tau) \right]^2 \right]^{1/2} \tag{22b}$$

or

Postulate III*: Moon, Spencer, Moon [34] - 1990

$$(r)_{III} = \left[[x(t) - \xi(t)]^2 + [y(t) - \eta(t)]^2 + [z(t) - \zeta(t)]^2 \right]^{1/2} \tag{22c}$$

Postulate II has been eliminated by study of the binary stars [35] unless we wish to assume a Riemannian space. For a Euclidean space only Postulates I* and III* are tenable.

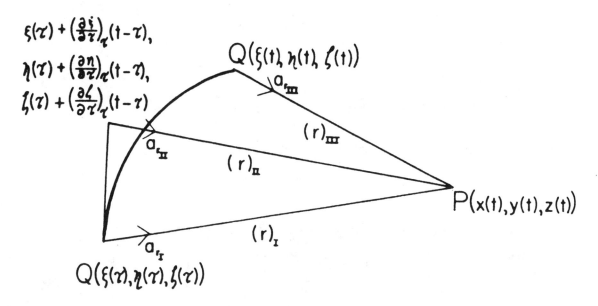

Figure 4
The distances $(r)_I$, $(r)_{II}$ and $(r)_{III}$ which occur in three postulates on the velocity of light.

At low velocities, $u, v \ll c$, all of the possible formulations of the force between moving charges reduce to only two. Suppose that the field is produced by a charge Q moving at velocity $\vec{v}$. The test charge at point P moves at velocity $\vec{u}$ and the force per unit charge on it is $\vec{\mathfrak{F}}$. According to classical theory using Equations (10), (18a) or (18b), (19a) or (19b), (20a) and Postulate I*,

$$\vec{\mathfrak{F}}_C = \frac{Q}{4\pi\epsilon_0(r)^2}\left[\vec{a}_r - \frac{\left(\vec{v}(t)\cdot\vec{a}_r\right)\vec{v}(t)}{c^2} + \frac{\vec{u}(t)\times\left(\vec{v}(t)\times\vec{a}_r\right)}{c^2}\right]$$

$$-\frac{Q}{4\pi\epsilon_0 c^2(r)}\left[\frac{d\vec{v}(t)}{dt} - \frac{\vec{u}(t)}{c}\times\left[\frac{d\vec{v}(t)}{dt}\times\vec{a}_r\right]\right].$$

(23a)

However, any of the Gaussian postulates at low velocities using Equations (10), (18a), (19c), (19d) or (19e), (20b) or (20c) and Postulate III* reduce to the relative velocity expression

$$\vec{\mathfrak{F}}_R = \frac{Q}{4\pi\epsilon_0(r)^2}\left[\vec{a}_r - \frac{\left(\left(\vec{v}(t)-\vec{u}(t)\right)\cdot\vec{a}_r\right)\left(\vec{v}(t)-\vec{u}(t)\right)}{2c^2} - \frac{\left(\vec{v}(t)-\vec{u}(t)\right)\times\left(\left(\vec{v}(t)-\vec{u}(t)\right)\times\vec{a}_r\right)}{2c^2}\right]$$

$$-\frac{Q}{8\pi\epsilon_0 c^2(r)}\left[\frac{d\vec{v}(t)}{dt} - \frac{d\vec{u}(t)}{dt} + \frac{\left(\vec{v}(t)-\vec{u}(t)\right)}{c}\times\left[\left[\frac{d\vec{v}(t)}{dt} - \frac{d\vec{u}(t)}{dt}\right]\times\vec{a}_r\right]\right].$$

(23b)

The Edwards [36] effect cannot be explained by the classical equation, but is explained by the relative velocity formulation [37]. Thus, there is some experimental evidence that the classical formulation should be replaced by the relative velocity formulation.

Expressions for the force between current elements can also be derived at low velocity. Again there are two possible formulations. For the classical theory the force between current elements, Figure 1, is

$$\left[d^2\vec{F}_C = \frac{1}{4\pi\epsilon_0 c^2}\left[\frac{I_1 I_2}{r^2} + \frac{I_2\frac{dI_1}{dt}}{cr}\right]\vec{ds}_2\times\left(\vec{ds}_1\times\vec{a}_r\right)\right]$$

(24a)

and

$$d^2\vec{F}_R = \frac{I_1 I_2}{8\pi\epsilon_0 c^2 r^2}\left[\vec{ds}_1\times\left(\vec{ds}_2\times\vec{a}_r\right) + \vec{ds}_2\times\left(\vec{ds}_1\times\vec{a}_r\right) + \left(\vec{ds}_1\cdot\vec{a}_r\right)\vec{ds}_2 + \left(\vec{ds}_2\cdot\vec{a}_r\right)\vec{ds}_1\right]$$

$$+\frac{1}{8\pi\epsilon_0 c^3 r}\left[I_1\frac{dI_2}{dt}\vec{ds}_1\times\left(\vec{ds}_2\times\vec{a}_r\right) + I_2\frac{dI_1}{dt}\vec{ds}_2\times\left(\vec{ds}_1\times\vec{a}_r\right)\right].$$

(24b)

Many experiments appear to show that the tangential forces predicted by the relative velocity formulation actually occur. Notable is the hairpin experiment performed by Ampère and de la Rive in 1822. A comparative analysis by Saria and Spencer [38] shows that the classical theory is inconsistent with experiment but that the relative velocity formulation appears to be correct.

Before definitive conclusions can be drawn on which of the 240 possible expressions for $\vec{\mathfrak{F}}$ derived from the postulational method is correct, all of the pertinent experiments must be analyzed carefully. The preliminary conclusions drawn from the analysis of a small number of experiments would lean toward the conclusion that a Gaussian type formulation is preferable to the classical formulation. If the relative velocity formulation is employed and the study of transformation equations is generalized to accelerated coordinate systems [39] then holor transformations must be employed instead of tensor transformations.

References

1. Newton, I., *Philosophiae Naturalis Principia Mathematica*, S. Pepys, London, 1686.

2. Lagrange, J. L., *Mém. de Berlin*, 1777.

3. Laplace, P. S., *Mém. de l'Acad*, 1785, p. 113.

4. Coulomb, C. A., *Mém. de l'Acad*, 1785.

5. Poisson, S. D., *Bull de la Soc. Philomathique*, Vol. 3, 1813, p. 388.

6. Poisson, S. D., "Mémoire sur la théorie du magnétisme," *Mém. de l'Acad*, Vol. 5, 1824, p. 247.

7. Oersted, H. C., *Experimenta circa conflictus electrici in acum magneticam*, Copenhagen, 1820.

8. Ampère, A. M., "Mémoire sur la théorie mathématique des phénomènes électrodynamiques, uniquement déduite de l'expérience," *Mém. de l'Acad. des Sci.*, Vol. VI, 1823, p. 175.

9. Maxwell, J. C., *A Treatise on Electricity and Magnetism*, Vol. II, Third Ed., Clarendon Press, Oxford, 1904, p. 175.

10. Gauss, K. F., "Zur mathematischen Theorie der elektrodynamischen Wirkung," Werke (Göttingen, 1867), Vol. V, p. 602.

11. Gauss, K. F., Werke, p. 627.

12. Weber, W., "Elektrodynamische Maassbestimmungen," *Ann. der Physik and Chemie*, Vol. 73, No. 2, 1848, p. 193.

13. Wesley, J. P., "Weber electrodynamics extended to include radiation," *Speculations in Science and Technology*, Vol. 10, No. 1, 1986, p. 47.

14. Pappas, P. T., "The original Ampère force and Biot-Savart and Lorentz forces," *Nuovo Cimento*, Vol. 76B, 1983, p. 189.

15. Phipps, T. E. Jr., *Heretical Verities: Mathematical Themes in Physical Description*, Classic Non-fiction Library, Urbana, Ill., 1986; T. E. Phipps and T. E. Phipps, Jr., "Observation of Ampère forces in Mercury," to be published.

16. Graneau, P., *Ampère-Neumann Electrodynamics of Metals*, Hadronic Press, Nonantum, Mass., 1985.

17. Assis, A. K. T., "On Mach's Principle," *Found. Phys. Lett.*, Vol. 2, 1989, p. 301.

18. Faraday, M., "Historical sketch of electromagnetism," *Ann. Phil.*, Vol. 2, 1821, p. 195.

19. Faraday, M., *Experimental Researches in Electricity*, London, 1844, 1849, 1855.

20. Neumann, F. E., "Die mathematischen Gesetze der inductirten elektrischen Stroeme," *Akademie der Wissenschaften*, Berlin, 1845, p. 1.

21. Thomson, W., *Camb. and Dub. Math. Jour.*, Vol. 2, 1847, p. 61.

22. Maxwell, J. C., *Maxwell's scientific papers*, Vol. I, II, Cambridge, 1890.

23. Maxwell, J. C., *Treatise on electricity and magnetism*, Vols. I and II, Oxford, 1873.

24. Lorenz, L., Oversigt over det. K. danske Vid. Selskaps Forhandlinger, 1867, p. 26; *Ann. d. Phys.* Vol. 81, 1867, p. 243; *Phil. Mag.*, Vol. 34, 1867, p. 287.

25. Whittaker, E., *A History of the Theories of Aether and Electricity*, Philosophical Library, New York, 1951, p. 310.

26. Lorentz, H. A., *The Theory of Electrons*, G. E. Stechert & Co., New York, 1906.

27. Hertz, H. R., *Electric Waves*, Tenbner, Leipzig, 1892; Dover, New York, 1962.

28. Bush, V., "The force between moving charges," *J. Math. Phys.*, Vol. 5, No. 3, 1926, p. 129.

29. Moon, P. and D. E. Spencer, "Electromagnetism without magnetism: An historical sketch," *Amer. J. Phys.*, Vol. 22, 1954, p. 210; "Interpretation of the Ampère force", *J. Franklin Inst.*, Vol. 257, 1954, p. 203; "The Coulomb force and the Ampère force", *J. Franklin Inst.*, Vol. 257, 1954, p. 305; "A new electrodynamics," *J. Franklin Inst.*, Vol. 257, 1954, p. 369.

30. Mirchandaney, A. S., D. E. Spencer, S. Y. Uma and P. J. Mann, "The theory of retarded potentials," *Hadronic J.*, Vol. 11, 1988, p. 231; "The electromagnetic fields of a moving charge," to be published in *Hadronic J. Supplement*.

31. Liénard, A., "Champ électrique et magnétique produit par une charge électrique concentrée en un point et animée d'un movement quelconque," *l'Eclairage Electrique*, Vol. 16, 1898, p. 5, 53, 106; E. Wiechert, "Elektrodynamische Elementargesetze," *Ann. d. Phys.*, 4, 1901, p. 680.

32. Einstein, A., "Uber das Relativitatsprinzip und die aus demeselben gezogenen Folgerungen," *Jahrbuch der Radioaktivitat*, IV, p. 422-462; V, p. 98-99 (Berichtigungen), 1907.

33. Ritz, W., "Recherches critiques sur l'électrodynamique générale," *Ann. Chim. Phys.*, Vol. 13, 1908, p. 1451.

34. Moon, P., D. E. Spencer, and E. E. Moon, "Universal time and the velocity of light," *Physics Essays*, Vol. 2, 1989, p. 368.

35. Moon, P., D. E. Spencer, and E. E. Moon, "Binary stars from three viewpoints," *Physics Essays*, Vol. 2, 1989, p. 275.

36. Edwards, W. F., C. S. Kenyon and D. K. Lemon, "Continuing investigation into possible electric fields arising from steady conduction currents," *Phys. Rev. D*, 14, p. 922, (1976).

37. Spencer, D. E., S. Y. Uma and P. J. Mann, "The force between moving charges and the Edwards effect," *The Thorny Way of Truth*, International Publishers, East and West, Part IV, Graz, Austria, 1989.

38. Saria, S. M. and D. E. Spencer, "Three interpretations of the hairpin experiment of Ampère and de la Rive," to be published.

39. Uma, S. Y. and D. E. Spencer, "Transformation equations in accelerated coordinate systems", to be published in *Physics Essays*.

40. Moon, P. and D. E. Spencer, *Theory of Holors*, Cambridge University Press, Cambridge, England, 1986.

From Howard Eves to *Africa Counts* to Multicultural Mathematics Education

Claudia Zaslavsky
45 Fairview Avenue, #13-I
New York, NY 10040

Howard Eves is responsible for my writing *Africa Counts: Number and Pattern in African Culture*. Furthermore, the publication of this book led to an entirely new direction in my life as a mathematics educator.

For years I had been a conventional teacher, using conventional textbooks, and imparting conventional high school mathematical wisdom in a way that I hope made at least a little sense to my students. I was fortunate to be teaching in a district, just north of New York City, that was known nationwide for its decision in 1951 to integrate its schools by busing. In the sixties some students requested courses in African history and Swahili. The faculty, too, was offered an optional course in African history. For my term paper in the course I decided to research the development of mathematics in Africa south of the Sahara, but was amazed to discover that there were no books or articles listed in any library catalogs under "African mathematics." In fact, there was no such heading in library card catalogs. I faced a real challenge!

Mathematical practices and concepts arise out of the real needs and interests of people in all societies, in all parts of the world, in all eras of time. All peoples have invented mathematical ideas to deal with activities such as counting, measuring, locating, designing, and, yes, playing, with corresponding vocabulary and symbols to communicate their ideas to others [1]. I was convinced that African peoples were no exception, and that such information must be available, but how would I locate it? I decided to write a form letter requesting information about the subject that I later called "African sociomathematics": numbers and numeration systems, applications of mathematics to measurement, record-keeping, and trade; geometry of architecture; form and pattern in art; and games of skill and of chance. I sent this letter to authorities in African studies and history of mathematics, both in the United States and abroad.

Among the authorities was Howard Eves, then editor of the "Historically Speaking" section of the *Mathematics Teacher*. He responded, in his delicate handwriting, that he had no information on the subject, but that if I were successful in gathering material, he would consider publishing an article on the subject. Eventually I was successful, and the article appeared in April, 1970 [2].

At that time, Howard Eves was consulting editor for the Prindle, Weber and Schmidt Complementary Series of paperbacks "devoted to supplementary, auxiliary, and enrichment materials at all levels of mathematical education and instruction." I was delighted to be offered a contract to write a book of about one hundred pages for this series. Eves concluded his letter with this forward-looking statement; "As you know, a project like this would be very useful at this point of our country's history."

The possession of a contract to write a book provides excellent motivation to do research. Following the publication of the article in the *Mathematics Teacher*, letters began to pour in from many parts of the world, some suggesting further contacts and others recommending readings. I received several articles and a Nigerian student's dissertation on the mathematical practices of his people. Meanwhile I was gathering bits and pieces of information by skimming through hundreds of books and articles by historians, anthropologists, colonial officers, and missionaries, many written in the early twentieth century to better enable the European powers to maintain their domination over colonial Africa.

By this time my son Alan had graduated from college and was teaching mathematics in a community-based secondary school in Kenya as a participant in the Harvard African Volunteer Project, a program similar to the Peace Corps. but with greater control by the host countries over the activities of the volunteers. Naturally, I planned a trip to Africa with the double purpose of visiting Alan and carrying on research.

In the summer of 1970 I carried out this plan. My first stop was London and the International African Institute, which was unique in cataloguing articles as well as books. In response to my form letter, the librarian had sent me copies of several cards in their catalog, enough to convince me that a few days at the Institute would be well spent. Then I proceeded to Kenya. Through a contact at the University of Nairobi, I arranged to interview representatives of several ethnic groups in that country. The university library was an excellent source of journals that were unavailable in New York. Alan's students and fellow teachers, people of various ethnic backgrounds, furnished lively anecdotes and allowed themselves to be photographed in appropriate poses to illustrate specific methods of finger counting peculiar to their own group.

Back in New York City, I contacted several African doctoral candidates, who were only too happy to offer whatever information they had available. After all, who had ever displayed interest in the mathematical practices of their people? In school they had learned about Greece, Rome, and England, but nothing about their own lands and customs. In fact, they had been taught to denigrate everything African as having little worth. It was only after independence that African countries embarked on programs to reclaim their heritage. However, such programs rarely extended to mathematics. Europe and the United States provided the models for school mathematics in the 1970s and earlier.

More than to anyone, I am indebted to University of Wisconsin professor Donald W. Crowe for his tremendous contribution to my book. As a participant in the early 1960s in a United States sponsored group organized to fashion a modern mathematics curriculum for African countries, he had suggested ways of using local materials and knowledge, including games, as mathematical teaching aids. Although the group rejected his contributions as too unsophisticated, I was subsequently their fortunate recipient. While I was preparing my book, Crowe used a semester in

London to research topics in African mathematics. I was the beneficiary of examples of graph theory accompanying Angolan creation myths and inspired by Bakuba (Zaire) weaving, and of traditional patterns in cloth that embodied principles of transformation geometry. Crowe wrote a section on the latter topic for the book, selected suitable photographs from the British Museum collections, and helped in innumerable other ways.

Prior to the African trip, Prindle, Weber and Schmidt had requested an article on African mathematics for publication in their new journal for two-year colleges. This appeared in Fall, 1990 [3], and brought additional contributions of material.

At last I faced the formidable task of putting it all together. Family and friends were mobilized to help. My husband, Sam, had taken wonderful photographs in Africa, and our older son, Tom, proved to be a most careful editor. Various friends translated from German, French, and Portuguese, read and edited portions of the manuscript, and assisted in other ways. Appropriately, our accountant supplied the title "Africa Counts."

At last the job was done. Instead of a one-hundred page essay, I had several hundred pages of manuscript, plus many photographs, maps, and diagrams. Moreover, the Complementary Series of supplementary paperback mathematics books had been abandoned. *Africa Counts: Number and Pattern in African Culture* [4], a hardcover production, was to stand virtually alone in a sea of college mathematics textbooks, with only Howard Eves' wonderful collections of anecdotes [5] to keep it company.

Before the book appeared in print, Howard Eves performed another important service by carefully reading the manuscript. Not only did he examine the mathematical content and the style, but he also corrected my spelling. I had used so many sources by British-educated authors that I had made the error of writing "travelled" and "travelling" instead of "traveled" and "traveling."

Eventually Lawrence Hill agreed to publish a paperback edition [6], still in print today, although the hardcover version has been discontinued. In fact, in the past year the sales have increased appreciably.

Many of the mathematical practices described in *Africa Counts* can be adapted readily to all levels of the curriculum. Initially the students in my own school district were the subjects of my experimental curriculum units [7]. Eventually the book became a mainstay of college courses involving "non-Western" cultures. Meanwhile I had extended my own scope to include the mathematical practices and games of many peoples, as exemplified by [11], [12], and [13], among others.

The idea is finally gaining acceptance that the mathematics developed by peoples outside of the mainstream is a field worthy of study. The International Study Group on Ethnomathematics [14] was formed in the mid-1980s, and the African Mathematical Union's Commission on the History of Mathematics in Africa [15] soon thereafter. Recently published books in this area include [16] and [17].

REFERENCES

1. Bishop, A. J., *Mathematical Enculturation*, Kluwer, Dordrecht, 1988.

2. Zaslavsky, C., "Black African Traditional Mathematics," *Mathematics Teacher*, 63 (1970), 345-356.

3. Zaslavsky, C., "Mathematics of the Yoruba People and of Their Neighbors in Southern Nigeria," *The Two Year College Mathematics Journal*, 1 (1970), 76-99.

4. Zaslavsky, C., *Africa Counts: Number and Pattern in African Culture*, Prindle, Weber & Schmidt, Boston, 1973.

5. Eves, H., *In Mathematical Circles*, et al., Prindle, Weber & Schmidt, Boston, 1969.

6. Zaslavsky, C., *Africa Counts*, Lawrence Hill Books, Brooklyn, NY, 1979.

7. Zaslavsky, C., "Multicultural Mathematics Education for the Middle Grades," *Arithmetic Teacher*, 38 (February, 1991), 8-13.

8. Campbell, P. J., "An Experimental Course on African Mathematics," *Historia Mathematica*, 3 (1976), 477-478.

9. Zaslavsky, C., "Teacher Education: Mathematics in Other Cultures," *Historia Mathematica*, 6 (1979), 189-194.

10. Ascher, M., "Mathematical Ideas in Non-Western Cultures," *Historia Mathematica*, 11 (1984), 76-80.

11. Zaslavsky, C., "Bringing the World into the Math Class," *Curriculum Review*, 24, (January 1985), 62-65.

12. Zaslavsky, C., *Multicultural Mathematics: Interdisciplinary Cooperative-Learning Activities*, J. Weston Walch, Portland, Maine, 1993.

13. Zaslavsky, C., "Symmetry in American Folk Art," *Arithmetic Teacher*, 38 (Sept. 1990), 6-12.

14. The president is Dr. Gloria Gilmer, Math-Tech Inc., 9155 North 70th Street, Milwaukee, Wisconsin 53223.

15. The chair is Dr. Paulus Gerdes, C.P. 915, Maputo, Mozambique.

16. Washburn, D. K. & D. W. Crowe, *Symmetries of Culture*, Univ. of Washington Press, 1989.

17. Ascher, M., *Ethnomathematics*, Brooks/Cole, Belmont, California, 1991.

An Alternative to the Conventional Wisdom of a Term Paper in History of Mathematics Survey Courses

Joe Albree
Auburn University at Montgomery
Montgomery, AL 36117-3596

I bought my first copy of an older edition of Howard Eves's *An Introduction to the History of Mathematics* at a second-hand bookstore when I was a beginning graduate student and full of the idealism of a late 1960s Teaching Assistant. When I dove into the book, I quickly discovered its most distinctive feature, the unequalled Problem Studies. And the more I worked and explored, the richer these studies became. The tip-off to their genius came in the Introduction.

> "Some of the Problem Studies concern themselves with historically important problems and procedures; ... and many can lead to short 'junior' research papers by the student."

This is what mathematics and its teaching really is! Because mathematics is man's oldest science, it should be explained. The stories of the roots and the twists and the turns are just as much a part of a complete understanding of a theorem or a theory as the technical details of the modern definitions and proofs. How did we progress from Newton's fluxions to the calculus I had learned in engineering school and was now teaching? There was a research paper, perhaps a little more than "junior," but missing from my graduate training. These are the kinds of challenges that a real student should find exhilarating. Sure, why not have term papers and 'junior' research – even in calculus class!

As a graduate student, I did not have the authority to destroy the syllabus of a multi-sectioned calculus course to accommodate anything as radical as a term paper. But, I was determined that, one day, I would teach a course in which the student would have this opportunity. My chance came when I began teaching a survey course in the history of mathematics using Howard Eves's book. The chapter bibliographies and the Problem Studies were fertile points of departure for term papers and other student projects. However, not all of my students caught the excitement of "junior" research. Over the years, I have devised an alternative to the standard term paper which, I believe, accomplishes at least most of its ambitious aims. In this presentation I describe briefly some of the problems that I have encountered with term papers, and some of the procedures that I have devised in my efforts to help my students with their exploration of mathematical concepts.

PROBLEMS

My misgivings with the standard term paper were reinforced at the HPM (History and Pedagogy of Mathematics) meeting held at the Joint Meetings in New Orleans in January 1986. It seemed that everyone had complaints about poor student performance on term paper assignments in history of mathematics courses. We had responded to the challenge of Howard Eves's "junior" research, but we had also been disappointed with the poor results of our students' work.

As I listened to the manifold and awful complaints and I reflected on my own experiences, I realized that universal condemnation of student term papers was not a correct description of the phenomenon with which we were wrestling. It was much more complex. A couple of my students' works have been turned into papers published in regional mathematics journals [1], [17]. There was "good news" and there was "bad news."

The key was almost trite: of course, some students did better than others. The essence of the problem was that each time I taught my course, some of the works produced by the students were well off the curve, some at the low end and some at the upper end. This recognition has prompted me to try to define more carefully some of the problems which my students have had in attempting to meet Howard Eves's challenge. The following, from my experience, is a general list of the sins committed by students in the name of the term paper in a survey history of mathematics course.

A.   **Inappropriate topic:** There was the student who tried to tell the whole story of the history of cartography (not just of the geometry of map projections) in 8 to 12 double-spaced typewritten pages. Another student wanted to stretch the history of the minus sign into a paper of this length. Each one missed the mark badly. In my own defense, there was a limit to the amount of counseling an instructor could provide in helping a student define the scope of his/her work.

Another way in which to describe this problem is to acknowledge the tension that exists between breadth and depth. We must take special care not to extinguish the passion of the student who wants to write a paper on the "history of everything" or to diminish the ardor of the student who desires to know everything there is to know about topic X, no matter how narrow topic X may be. On the other hand, there are practical limits on the length of the quarter or semester, the library's resources, and the student's and the instructor's time and knowledge.

B.   **Mathematical Content:** Mathematical content may be our dirty little secret. Some of the weaker students see the term paper as an opportunity to fill up 8 to 12 pages with "facts", biographical or historical or even fanciful, but disingenously avoid any meaningful mathematics, and yet still expect to fulfill at least the letter of the assignment. On the other hand, I had a student who thought he could write the history of algebraic semigroups by just quoting theorems and proofs and appending the names and dates of their authors.

C.   **Variablility in the Use of Sources:** One of the most telling observations, as I contemplated the HPM discussions and reviewed the successes and failures of my history of mathematics students, was the variability in the use of sources. Many had not done much digging. Too many had just used the first book or two they encountered on their topic, and one or two had not even progressed much beyond general encyclopedias. There were a couple of notable exceptions, but I was surprised at the number of students who were unimaginative and unresourceful in pursuing a topic, who lacked a scent for the chase that I thought "junior" research would inspire.

D. **Cut and Paste Work:** Unfortunately, when they had located a source, a few of my students could not extract the information and put it into their own words. Theirs was what I call cut and paste work. The student would copy whole phrases, sometimes even a sentence or two, change one or two insignificant words somewhere in the middle, and then pass it off as his/her own work. One cannot fail to notice this practice when grading such papers because the writing is markedly disjointed. Some students were just lazy, but a few genuinely did not know any better.

E. **Writing Mechanics:** Most unsettling of all, I have had a small number of students whose writing mechanics were deficient. Some students could not use pronouns correctly or couldn't be counted on to make their subjects and verbs agree in tense and number, while others exhibited inconsistencies in the content of their writing. This has been profoundly disappointing, of course. But, I must hasten to add that I have also had perhaps an equal number of student writers who were quite lucid, who explained some good mathematics with not only clarity but imagination and insight.

This is quite a lengthy list of ills, and let me say again, the variability of the performances has been quite striking to me. "The Closing of the American Mind" may be a catchy phrase to sell books, but it connotes a uniformity of malaise which I think misses the reality of our predicament.

REFINE OUR GOALS

This list prompted me to stand back and re-examine my goals for "junior" research; is there any other way we can reach for these objectives, or at least most of them? Of most importance to me is breaking the bonds of the textbook with which nearly every mathematics course is shackled (I exclude the so-called "Moore Method" courses).

I think this is the heart of Eves's invitation to do "junior" research. And, his Problem Studies are truly revolutionary. After all, the "exercises" in the Rhind Papyrus were arranged according to increasing difficulty; and so they are in every other mathematics textbook I know of for the past 3600 years - except for Eves's *History of Mathematics*. This is culture shock to students, and it's just what they need to begin this emancipation from the "sacred" textbook. With problems organized around an issue or theme, problems which provoke discussions and connections and even reveal ambiguities, the textbook is not the last word, it's only the first!

I would like each student to learn to be comfortable and confident in using some basic library research skills and tools. For the history of mathematics, this means becoming familiar with the *Dictionary of Scientific Biography* [13], knowing how to use bibliographies such as Kenneth May's [16], and Joseph Dauben's [5], and being familiar with *Historia Mathematica* [15] and the *Archive for History of Exact Sciences* [2]. I really want more, especially for the teachers. When their students ask irreverent questions, I want these teachers to have the skills, resourcefulness, and tenacity to dig to the bottom.

Organizational and writing skills are a must. Writing taught in English classes tends to be about literary matters, and naturally so because this is what English teachers know best. Writing about mathematics can be hard. Students need some good role models, Eves's *History* is excellent, and so are many of the items a student with a little persistence will find in the library. The student must do some writing. No one learns how to swim without getting wet.

There is a certain experiential knowledge which comes from successfully completing a major piece of writing. I don't want to oversell a term paper for just one course. But I do want a piece of work of which the student can be proud. The student met the challenges, and with the completion of the work, it should be clearly the student's possession. If the term paper is not working in a survey history of mathematics course, then what else can we do?

INTRODUCTION AND ORIENTATION

Samuel Johnson said it best, "Knowledge is of two kinds; we know a subject ourselves, or we know where we can find information upon it." I want my students to be a little disoriented on the first day of my class, they should know that whatever this history of mathematics is, it is not just another ticket to be punched. So, on the first day of class, we go to the library and we go through my ten page *Annotated Bibliography* for the course. The idea of making such a bibliography was generously suggested by Barnabas Hughes when we exchanged notes about our respective courses the first time I taught history of mathematics back in the late 1970s. My bibliographic effort is broken down into the following sections.

I. Bibliographies and references (for example [5] and [16])

II. Books on the history of mathematics
    A. General histories of mathematics
    B. Some specific topics (a highly selective list, for example [4] and [14])
    C. Biographies (a highly selective list, for example [7] and [8])
    D. Anthologies, collections, and source books

III. Journals
    A. Historical
    B. Others like the *Monthly* [21] and the *Intelligencer* [23] which include historical articles

IV. History of Mathematics in America

Part of each entry in this bibliography is a special place where the LC number is either included or can be written by the student the first time he/she looks up and/or uses that work. It is my design that students will use this bibliography throughout the course, including when they work on Eves's Problem Studies (and they must footnote appropriately), not just for the "writing" assignments. I very much hope all of this material will be of use to students long after the course is over.

PROGRESSION OF LITERATURE REPORTS

To replace the term paper, I have devised the following progression of Literature Reports (LRs). The student is simply to report, in approximately two pages, on what he/she has read from each of the sources. Here is the progression, with an example of an actual student selection:

A.   A chapter or a section from a survey history of mathematics book - decimal fractions, from Smith [19], pp. 235-247;

B.   A chapter or a section from a book on the history of a mathematical topic or time period, or a biography of a mathematician - the four arithmetic operations, from Gillings [12], Chapter 3, pp. 11-23;

C.   A journal article - the separation of algebra from arithmetic, in Dubbey [9];

D.   A primary source - letters between Fermat and Pascal on the foundations of probability, in Smith [18], pp. 546-554;

E.   A primary source - solving first and second degree equations in Euler [11], Section IV, Chapters 1, 2, and 3, pp.186-206.

The idea behind this particular progression, is to lead the student away from the textbook and back to primary sources. This is the "emancipation" feature of "junior" research, where we break out of the confines of the holy textbook. It is truly ironic that in pursuing this objective, we are turning away from one of the best textbooks in all of mathematics; this, in fact, is one of Eves's great virtues! For the purpose of my course, I take a very liberal view of "Primary Sources," since the great majority of American college students today are not proficient in a language other than English. But, for example, G. J. Toomer's *Diocles on Burning Mirrors* [24] is excellent; even excerpts from one of the *Source Books* [3], [18], and [20] are acceptable.

Let me expand on some of what I call (for want of another name) the "LR Ground Rules". The student must choose his/her topic at least a week before the due date of that LR, and I keep a check list in my course file. This allows me to offer some counseling on where the student might start looking, breadth and depth questions, etc. This also helps us keep the student honest, in that he/she cannot just write an LR on the first item he/she comes across in the library. The student may choose the same topic for all five LR's and thereby perhaps even string them together into the beginning of a "junior" research project, or the student may just shop around as his/her interests are aroused by different parts of the course.

I encourage the use of word processors. Equations and mathematical notations may be carefully printed in by hand for students who do not use TeX. The title should just be a correct (*Chicago Manual of Style* [22] or Turabian [25]) bibliographical citation of the work that is the subject of the LR.

In all LRs after the first, the student must attach a worksheet that consists mainly of the library trail. This is a piece of notebook paper, handwritten, in pencil, a diary of sorts listing the references the student used (usually starting with either Eves' [10] or May's *Bibliography* [16], or some other bibliography, or the *DSB* [13]). If and when a student alters or refines his/her topic, that must be recorded also. As can be seen, this worksheet is a substitute for at least some of the persistence that a student should learn in the course of producing a good term paper. There must be non-trivial mathematical content. Writing mechanics are strict.

CONCLUSION

What does this progression accomplish? On the scale of cognitive traits and attributes, analysis is one of the higher achievements, and synthesis is usually considered to be the next step up. In giving up the term paper, I concede that I have lost synthesis as a goal for my students. By the nature of my Literature Reports, there is no synthesis. This is a dear price to pay. What, if anything, have we gained for this extravagant cost?

The appropriateness of the topic problem is neutralized. The student who chooses a subject which is too grandiose won't be overwhelmed. The weak student shouldn't be intimidated, and if the topic is trivial the student will be quickly found out (or, better yet, will find out for himself/herself). Most of the goals of "junior" research have been met, with the major exception of cultivating the student's power of synthesis, of course.

The feedback has given us a good deal. Problems with writing mechanics and the minority of students who want to avoid any challenging mathematics are nipped in the bud. Abilities and resourcefulness with research tools and skills are considerably helped. Students must demonstrate, especially through their work sheets and library trails, that they know their way around some bibliographies, journals, and some of the literature. The student's skills and abilities of analysis are improved with the continued practice and the increasing depth of our progression.

I think the greatest benefit is that every student gains a sure knowledge of what primary sources are. Most of the best of my former term paper writers depended almost exclusively on secondary sources. Now, every student must wrestle with some of the harsh realities of the "real McCoys," from elongated S's, to unfamiliar and archaic notation, to different terminologies, to idiosyncracies and even sometimes errors or misconceptions. (I cannot resist one of my favorites. In the Second Book of his *la Geometrie* [6, p. 95], Descartes says

> "...I shall have given here a sufficient introduction to the study of curves when I have given a general method of drawing a straight line making right angles with a curve at an arbitrarily chosen point upon it. And I dare say that this is not only the most useful and most general problem in geometry that I know, but even that I have ever desired to know."

Mais non, Rene! It's the derivative which is the most useful and most general problem in geometry, not the normal!)

These can be priceless experiences for students. I know there is no free lunch. But I do believe that for a survey course in the history of mathematics, replacing the term paper with this progression of literature reports has produced both practical skills and real intellectual growth for my students.

References

[1] Albree, J. and P. Johnston, "Review of *The Rhind Mathematical Papyrus* by Arnold Buffum Chase," *Alabama Journal of Mathematics* 7, no. 1 (1983), 15-20.

[2] *Archive for History of Exact Sciences*, Berlin: Springer Verlag, 1960 - .

[3] Birkhoff, G., editor, *A Source Book in Classical Analysis* Cambridge, MA: Harvard University Press, 1973.

[4] Cajori, F., *A History of Mathematical Notations*, 2 volumes, LaSalle, IL: Open Court, 1928-1929.

[5] Dauben, J. W., *The History of Mathematics from Antiquity to the Present: A Selective Bibliography*, New York: Garland, 1985.

[6] Descartes, R., *The Geometry of Rene Descartes*, translated from the French and Latin by D. E. Smith and M. L. Latham, New York : Dover, 1954.

[7] Dick, A., *Emmy Noether*, 1882-1935. Boston: Birkhauser, 1981.

[8] Dijksterhuis, E. J., *Archimedes*, Princeton, NJ: Princeton University Press, 1987.

[9] Dubbey, J. M., "Babbage, Peacock, and modern algebra," *Historia Mathematica*, 4 (1977), 295-302.

[10] Eves, H., *An Introduction to the History of Mathematics*, 5th edition, Philadelphia: Saunders, 1983.

[11] Euler, L., *Elements of Algebra*, Translated from the French; with the Notes of M. Bernoulli, &c and the Additions of M. de la Grange, 5th edition, London: Printed for Longman, Orme, and Co., 1840.

[12] Gillings, R. J., *Mathematics in the Time of the Pharaohs*, Cambridge, MA: MIT Press, 1972.

[13] Gillispie, C. C. (editor), *Dictionary of Scientific Biography*, 16 volumes and supplements, New York: Scribners, 1970-1980.

[14] Grabiner, J., *The Origins of Cauchy's Rigorous Calculus*, Cambridge, MA: MIT Press, 1981.

[15]   *Historia Mathematica*, New York: Academic Press, 1974 - .

[16]   May, K., *Bibliography and Research Manual of the History of Mathematics*, Toronto: University of Toronto Press, 1973.

[17]   Moran, R. and J. Albree, "Construction of magic squares by an early Alabama mathematician," *Alabama Journal of Mathematics*, 4, no. 2 (1980), 22-30.

[18]   Smith, D. E., *A Source Book in Mathematics*, New York: Dover, 1959.

[19]   Smith, D. E., *History of Mathematics*, volume 2, Boston: Ginn and Co., 1953.

[20]   Struik, D. J., editor. *A Source Book in Mathematics, 1200 - 1800*, Cambridge, MA: Harvard University Press, 1969.

[21]   *The American Mathematical Monthly*, Washington, DC: MAA, 1894 - .

[22]   *The Chicago Manual of Style*, 13th edition, Chicago: University of Chicago Press, 1982.

[23]   *The Mathematical Intelligencer*, New York and Berlin: Springer Verlag, 1978 - .

[24]   Toomer, G. J., *Diocles On Burning Mirrors*, Berlin: Springer Verlag, 1976.

[25]   Turabian, K. L., *A Manual for Writers of Term Papers, Theses, and Dissertations*, 4th edition. Chicago: University of Chicago Press, 1973.

*DERIVE*ing a New Approach to Calculus

Steven M. Amgott
Widener University
Chester, PA 19013

Jane E. Friedman
University of San Diego
San Diego, CA 92037

In this paper we will describe a calculus sequence incorporating the computer algebra system *Derive*, implemented in Fall, 1990, at Widener University. The symbolic manipulation and numerical abilities of this system are used by students to supplement their basic skills and help them concentrate on independent problem-solving strategies, while its graphical component enhances their conceptual understanding. Students "discover" much of the mathematics with the aid of laboratories and work on more involved applied take-home projects. We believe that this approach can turn mathematics from a spectator sport into a participatory one.

Goals

Of the many goals which we had for the re-worked calculus sequence, increasing students' involvement in their own education was first and foremost. It is, of course, extremely unlikely that this can be done in the traditional lecture format. Much recent research into mathematics education supports the idea that lecturing may be one of the least effective ways to teach mathematics to our students.

Many students equate Calculus with the rote manipulation skills on which they spend much of their time. These manipulations often obscure the most meaningful ideas from the course. The advent of symbolic computation systems has made us question the emphasis these skills receive in our courses. We felt that the time was ripe to change the way we try to teach students, and in the process, change the calculus course itself.

We decided on an in-class laboratory format in which students would work in pairs on instructor-written laboratories designed to lead them to an understanding of basic concepts in the course. Many of these laboratories were given before any classroom discussion of the concepts occurred. We hoped that the students would form ideas based on their laboratory work, enabling them to be more involved in classroom discussions that followed. The instructor's role changed from that of being the source of knowledge to that of being a coach in the educational process. By allowing students to discover ideas on their own, we hoped to make the material more exciting and alive for them. Not only would this make the course more interesting for the students, but they would feel a sense of ownership toward the material since they "discovered" it themselves. We expected that this process would increase both their mastery and retention levels.

We identified several secondary goals for the project. We wanted an improvement in the conceptual understanding of our students, who usually only master techniques. Mathematics is much more a mode of thinking than a collection of facts. In this sequence of courses we hoped to begin the process of teaching students to think mathematically.

We hoped for an improvement in the ability of our students to solve more difficult multi-step problems. Too often, mathematics text problems can be solved quickly by following the appropriate paradigm example in the text. Students get the impression that all problems are solved either quickly, or not at all. They give up if it takes them more than five or ten minutes to get the answer. We also wanted to have them work on open-ended problems, so that they experienced the fact that not every problem has a unique solution.

We expected that use of the computer would also allow us to raise the realism level of applications. Realistic applications tend to lead to "messy" work and complications in calculations. Using the computer to do this work would free the students to concentrate on the basic ideas of the problem. We hoped that the students would be willing and persistent enough to work through longer problems, and would gain a sense of accomplishment in finishing the work.

A goal formulated somewhat later was to improve the ability of our students to write, especially about mathematics. Mathematics is not just about concepts, it is also about the communication of ideas to others. We feel that this last goal has been ignored far too long by mathematics instructors at all levels.

As at many schools, there is a high drop-out and failure rate in the calculus sequence at Widener, and we hoped the new approach would somewhat reduce this. Two features of the course in particular would contribute toward an improvement in the retention rate. By creating a more interesting and exciting course, we expected to increase the desire of the students to work and succeed in the class. By employing lab partners and project teams, we hoped that the students would learn from one another and provide mutual support which would increase the likelihood of their success in the course.

Structure and Organization

The course we are describing is only the first step on the road to a radically different Calculus course. Some aspects of the course remain traditional; others are novel. In the future we expect the course to evolve in less traditional directions. This in itself gives the students a valuable sense of Mathematics as an exciting and alive endeavor.

At Widener, only some of the sections of Calculus utilized this new approach, making it harder to break completely free of the old methods of teaching. We needed to make some adjustments to the standard syllabus, but were constrained to keep them to a minimum. We had not planned originally on using a text, but our engineering faculty felt very strongly that a text for reference was essential for their students. The textbook we used was Finney and Thomas, but any reasonable standard text would have worked. We assigned readings and some standard homework problems from the text, but they were not stressed. We expect that new texts will be marketed in the future which will be much more appropriate.

What was stressed were in-class laboratories which were designed to encourage discovery and experimentation. Students worked on these in pairs. We met in a computer equipped classroom three out of four meetings a week. This gave us great flexibility in scheduling the labs. The students began to work on the labs in class with the instructor circulating among them. In this way, we were able to insure that the students at least got started in the right direction. We tried to introduce as many concepts as possible through labs, thus allowing the students to experience a sense of discovery. *Derive* was well suited to combining symbolic, graphical, and numerical work, and many of the labs integrated all three approaches to a given problem. For each lab each pair of students submitted a written report, in which they described and generalized their results.

Students also worked in pairs on take-home projects, which were longer and more involved than the labs. We had hoped to make the projects applied in nature, and solicited ideas for realistic applications from our colleagues in other disciplines. We were not particularly satisfied with the projects we assigned and hope to develop better ones in the future.

The labs and projects figured significantly in the assignment of final grades to the students. In the grading of the labs and projects, we were somewhat concerned with writing quality, so we allowed the students to rewrite their work, and at times insisted upon it. This type of grading was very different from the kind of grading to which we were accustomed.

Our in-class exams contained both standard questions and deeper, more conceptual questions. In addition, each exam, including the final, had a take-home section. The take-home sections included problems which were more involved, both conceptually and computationally. We encouraged the student to use the computer when working on these problems.

Each semester we administered a "barrier" test to insure that the students could do basic computations. These will be renamed "mastery tests" in order to make them sound less threatening and to agree with current educational terminology. In the fall we tested them on derivatives, and in the spring, on integrals. The problems on these tests were standard and non-trivial, but also not of any great difficulty. The students were required to be nearly perfect, but were allowed to retake the exam as often as necessary to achieve this result.

Examples

The most important new idea in the course was the in-class laboratories. During the second class meeting, we presented the students with Laboratory 1. We felt that the best way to introduce the students to *Derive* was just to give the laboratories and to coach them as needed on the commands necessary to complete them. For the first laboratory, we included more details on the operation of *Derive* than became the norm. Although many never became expert *Derive*rs, the students seemed to have little difficulty in picking up the basic operations, with a little coaxing by the instructors.

The first laboratory contained much of the flavor of succeeding laboratories. First, it required the students to experiment with the graphical aspects of the program in order to form their own conjectures, not simply to test an idea given by the instructor. Second, it called for the students to generalize the conjecture they formed from one or more specific examples. Third, it required them to justify their conjecture by some form of mathematical reasoning, even if not quite a formal proof. As with all the laboratories, the instructors circulated in the classroom, helping students with the menu commands, and providing encouragement.

Laboratory 4 shows how we used the computer to introduce the basic course concepts. Given just after our work with limits, but before any discussion of derivatives, it was to stress the geometric picture of derivatives and tangent lines, and to introduce the idea of linear approximation. The derivative becomes the slope of the best linear approximation, rather than some strange limit calculated by hand. The students try to generalize the slopes they find to give a formula for the slope at any point on the curve. The exponential function makes an early appearance in our course, and this lab allows us to experiment with its derivative much earlier than would be possible in a standard course.

The second semester lab we included introduces the idea of Taylor polynomials. The only formal discussion preceding it was the idea of trying to create a polynomial whose derivatives of various orders have the same values at a single point as those of a given function. The graphs that the students draw in this lab help motivate the remainder term and the idea of convergence.

Other laboratories introduce the concept of a limit, numerically and graphically, provide work with numerical integration, introduce the Fundamental Theorem of Calculus, and work with infinite series and their partial sums.

In general we were less pleased with our projects. One which was somewhat successful involved student library research to obtain population data and estimates for the United States in this century and the world as far back as possible. These data are fitted informally to exponential functions, and the students report back on the correlation. They are asked to try to explain discrepancies in light of any historical events. A follow up to this project during the second semester used a logistic growth model to replace the exponential one. Since *Derive* provides the ability to work with differential equations numerically using the Euler method, an idea for the future is to incorporate some other topics usually saved for a differential equations course, such as predator-prey and pollution (mixing) problems. Other projects included more involved optimization problems with parameters, and an analysis of the hanging cable problem.

Exams were also changed from those given in the past. The inclusion of the routine differentiation problems on the "barrier" (mastery) tests allowed for some experimentation with the types of questions asked in class. Students used tables of values of functions to illustrate rules of differentiation and to perform numerical integration. From a plot of a function and its first two derivatives, students were to identify which was which. Another included interpreting statements about the economy and motion of a car in terms of derivatives of functions. Take-home sections included some numerical work and some work similar to labs and projects, but on which students worked singly. These exams have so far proven to be more difficult than regular exams, and modifications will have to be made in future years to make them less daunting to our students.

Student Response

One indirect measure of the students' feeling towards the course is the drop rate. In the fall these sections had a 0% drop rate, and in the spring a 4% rate. In comparison, the standard sections had drop rates of 25% in the fall and 16% in the spring. Considerably more than half of the students said they would both take the course again if they had it to do over and would recommend it to incoming freshmen, although often with a warning about the hard work involved. When asked if they wished to follow up on this experience by taking a third semester course using the computer, 12 said that they would like to continue, and two, that they would not. In addition, five students responded that they would have liked to have the opportunity, but were taking the course over the summer. Three did not need the course at all.

On the students' evaluations of the course, we received generally favorable comments about the use of the computer. Many felt that they had learned more than they would have otherwise; in fact, in the second semester, several students felt that we had not used the computer enough. One student was sure that knowledge of *Derive* would prove useful in the future. Most of the unfavorable comments were complaints about the amount of work involved. Some students also had legitimate complaints about the spacing of the work, and about the fact that our grading standards were sometimes unclear. These are areas in which we hope to improve the course in the future. Many of the students had positive comments about the group-work and felt good about the fact that grading was not based solely on in-class exams. Students have a great deal of anxiety about such exams and welcome a chance to show their abilities some other way. The majority of the students liked most aspects of the course.

Our Impressions

We were reasonably pleased with the outcome of these sections. The most gratifying aspect was the retention rate in the course, and the strong desire several students expressed to continue with a third semester course using *Derive* and other software. The goal of increasing student participation was definitely met; the group dynamics in the classroom were rivaled only in the past by honors classes and a few graduate classes taught by Dr. Amgott and by some math-education graduate courses taught by Dr. Friedman. There was also a marked improvement in the quality of writing in the work students submitted by the end of the second semester, even though no formal training was given to them. In future courses, we will try to make use of the University's Writing Center to help the students with their early reports and achieving even greater proficiency in their writing ability.

Although there was a reduction in the class time spent on routine work (e.g., methods of differentiation and integration), we do not feel that the students' abilities to do this work has suffered. This is especially true in regard to differentiation. The barrier tests sufficed to make the students hone these skills. We did put some standard problems on the first semester final to check differentiation skills, and the classes performed at least as well on these problems as regular sections we have taught in the past. Standard word problems seemed to become the new "routine" problems on our exams, and our students showed a great improvement in their abilities to handle them. In addition, most of the students did quite well on take-home involved problems with several parameters, even when they worked on them "by hand" rather than making the trip to the lab to use the computer to find the derivatives and solve the equations. Whether or not a good performance on routine problems is desirable pedagogically, it is clear that it is not necessary to spend a great deal of class time on them in order to achieve satisfactory performances by the students.

The jury is still out on whether or not the students showed an improvement in conceptual understanding. Part of the problem is that we do not have a good yardstick with which to measure, as the types of questions usually given on standard exams do not test this. Scores on the conceptual in-class exams we gave were not particularly high, but considering that the questions were not the types that students expect to see, they were not particularly low either. The students seemed to have a better grasp of the basic concepts, and were clearly able to handle questions that would have been considered very difficult by regular sections. Time, and their ability to integrate calculus concepts into their other courses, will tell whether or not we were successful in this endeavor.

To us, this experiment was a qualified success. Laboratories and (especially) projects need to be reworked and expanded, the latter with more input from faculty members in client disciplines. We will continue to search for more appropriate course materials, including a textbook that more fully embraces the experimental approach. Such books are in active preparation, and parts of them have begun to appear. We also expect to make further revisions in the course content, moving it further from the traditional course. The question as to whether the good student response can be maintained with larger sections will have to be addressed, as well as the need to get other faculty members at our institution involved. Teaching this type of course can be exhausting, time consuming, and at times quite difficult, but the payoff can be tremendously gratifying and even exhilarating.

References

[1]   *Calculus for a New Century*, MAA Notes 8 (1988).

[2]   Gilligan, L. and J. Marquardt, *Calculus and the Derive Program*, Gilmar Publishing Company, Cincinnati, 1990.

[3]   National Research Council, *Moving Beyond Myths - Revitalizing Undergraduate Mathematics*, National Academy Press, Washington, 1991.

Sample Laboratory Exercises with *Derive*

Math 141 - Laboratory #1:

In general, write equations necessary to shift the graph of any arbitrary function $y = f(x)$
  a)   up k units. What is the significance if k is a negative number?
  b)   to the right h units. What is the significance if h is a negative number?
  c)   up k units and at the same time right h units.
  d)   Give reasons that explain why your answer is true. The reasons should be convincing to someone who does not know anything about shifting graphs.

Math 141 - Laboratory #4:

Plot the function $y = x^2$.

  1)   The cross on the screen should be at the point $(1,1)$. If it is not, move it there. Center the screen on this point. Zoom in at least four times. (You may need to re-center occasionally.)
  2)   What seems to be happening to the portion of the graph being plotted?
  3)   Estimate the slope of the tangent line to the graph at the point $(1,1)$. You may use the cursor keys to move the cross to different points on the graph to aid in your estimation. What is the equation of the tangent line at $(1,1)$?
  4)   Check your equation by setting the scale back to 1 and 1, and plotting your equation of the tangent line. Zoom in a few times to see if the curve gets closer to the line or not.
  5)   Estimate an interval on the x-axis on which the y-coordinate of the tangent line is within .05 of the y-coordinate of the curve.
  6)   Do similar work to find tangent lines to the graph of the function $y = e^x$ at the point $(0,1)$ and $(1,e)$, and estimate the slope at as many points as necessary to guess a formula for the slope of the tangent line at the point (a,e^a).

Math 142 - Laboratory #5:

For each of the following functions, graph the function and the Taylor polynomials of the indicated degrees for the function centered at 0. Estimate the remainder terms $R_n(x)$ and $R_n(x)/x^n$ at several point approaching the given point. Describe what you see as the effect of changing n.
  1) $\sin(x)$ $n = 3, 6, 9,$ and 12
  2) $\ln(1+x)$ $n = 2, 3, 4, 5,$ and 6

Alternatives to the Lecture in Teaching

William Bonnice
University of New Hampshire
Durham, NH 03824

Many students don't want to be made responsible for their own learning and they resist new methods of teaching and learning. However, with encouragement and persistence they usually adapt, adopt, and, in the end, they are happy with the new experiences and the knowledge and maturity gained.

I have been experimenting with projects, cooperative learning in groups, and board presentations by the students ever since I started teaching twenty-nine years ago but only in the last five or six years have I incorporated these with other methods to the extent that now, in some courses, I do very little lecturing.

In this paper I would like to discuss mainly the methods the students and I used in a geometry course with Richard J. Trudeau's *The Non-Euclidean Revolution* (Birkhäuser Boston, 1987) as a text. Although Trudeau follows Book I of Euclid's *Elements* closely, even using the same numbering and proofs of Euclid, he points out where Euclid is found lacking from a modern point of view and he lists Euclid's tacit assumptions as additional axioms. Trudeau also avoids Euclid's use of superposition by adopting Side-Angle-Side as an axiom and therefore has to modify a few of Euclid's proofs.

After following Euclid through the proof of the Pythagorean Theorem, Trudeau moves beyond Euclid to a thorough development of the attempts which were made to prove Euclid's Postulate 5, the parallel postulate. He then goes on to replace Euclid's Postulate 5 by a strong version of the negation of Playfair's Postulate and uses that to develop hyperbolic geometry.

The Non-Euclidean Revolution is written from a historical perspective in an informal, thought-provoking style which lends itself to incorporating essay questions, writing, and other nonlecture methods into the course. The students really like "The Non-Euclidean Revolution" because "it is not written like a math text." I recommend it highly for a course in which to try out new methods.

Before I get into the details of the learning methods we used in this course I want to make an observation about the students' concern about grades. I used to despair because, whenever I wanted to try anything new, the first thing the students always wanted to know was how it would count toward their grade. Whether or not it was an effective way to learn seemed of little concern.

I finally realized that there is a bright side to this. Whenever you want the students to try anything new, tell them it will count toward a certain percentage of their grade. Be that percentage ever so slight, they will do it. But if an activity is made optional, no matter how beneficial it is, if it doesn't count toward their grade, pressure from other courses will keep students from carrying it out. Here is a listing of the topics that I will discuss.

I.	Getting to know one another.
II.	Getting started: syllabus and methods of learning.
III.	Individual work in class, followed by collaborative work in groups.
IV.	Board presentations.
V.	Journals.
VI.	Writing to learn.
VII.	Essay questions.
VIII.	Group projects.
IX.	Self-evaluations and teacher evaluations.
X.	Grading.
XI.	Student input.
XII.	Summary and observations.

I.   GETTING TO KNOW ONE ANOTHER

To have an informal class and to promote easy discussions, it is important at the beginning that everyone get to know one another and that everyone feels comfortable in the class. In the first class meeting, after briefly introducing the course and its contents, I tell something about the methods we will be using and point out that, to facilitate using these methods, it will be important for us to get to know one another and be comfortable with each other. Then we all sit in a circle facing each other. I begin by introducing myself and tell them some things about myself. Then we go around the circle and the students introduce themselves, tell where they are from, what they did last summer, why they are majoring in math, and what they hope to do after graduation.

In the first week I ask the students to write a "mathematical autobiography" telling their strengths and weaknesses in mathematics, about how their interest in mathematics developed, and what they hope to do with it. In this autobiography each student is to write a detailed description of at least one memorable mathematical event or experience, positive or negative, that they have had. When I have passed their autobiographies back, I ask for volunteers to read their experiences. Sometimes particularly interesting experiences would be embarrassing for a student to read. In this case I tell the story or experience myself without identifying the student who had the experience. Considerable time and effort is spent in the first few weeks to develop a class of students who know and care for each other.

II.   GETTING STARTED: SYLLABUS AND METHODS OF LEARNING

At the outset I want the students to get involved in deciding what they want to learn and how we are going to learn it. An early assignment is to look through the whole text and decide what material they'd like to learn, what they'd like to skip, what they thought was most interesting and important. I don't ask them for a detailed syllabus, but I do require that they write down a general outline of what they would like to learn and what is important to them. I also ask them whether they think it is important to go into things in depth and detail or whether they would prefer to cover more material in less detail. Setting a pattern for all their writing assignments, I ask them to give reasons for their preferences. If certain material must be covered in the course, I tell them what that is. Before they leave class to do this assignment, we have a discussion of the reasons for giving such an assignment.

As I read their papers I make a written summary of the responses. At the beginning of the next class, after passing back their papers, we have a discussion of what the course content should be, and by consensus (hopefully) we arrive at some guidelines for what will be covered in the course and at what level of detail. What I've just presented is typical of how I use writing assignments to generate lively class discussions.

After getting the course content out of the way, I tell them about my philosophy of teaching and the basis for it. I explain that, ideally, they will become responsible for their own learning, that they will become independent learners and will no longer need a teacher. A mentor, yes, but a teacher, no. I tell them that involvement and participation are the keys to learning, and then I talk about the various methods I propose for them to use and ask them to suggest reasons for each.

Again I give them a writing assignment, asking them to write a short comment on each of these methods, whether they think it ought to be used or not, and why. Finally, they should decide what percentage of their final grade each method should determine. As before, between classes, I read their papers and for my own benefit and for theirs I write a summary of their responses which I use at the next class as a basis for a discussion of what methods we are going to use in the course and what percentage their performance in each method will contribute toward their grade.

Some students enter this discussion wholeheartedly, whereas others just want me to make decisions and get on with the learning. This is typical of what happens when students are asked to take some responsibility for their own education. The role of the teacher is to try to bring about a consensus but not to let things drag out. At some point, the teacher must make decisions on what seems to be a consensus but inevitably some students are not satisfied. They will be given ample opportunity several times during the course to present their dissatisfaction verbally and in writing and, at times, even bring about changes during the semester.

III. INDIVIDUAL WORK IN CLASS FOLLOWED BY COLLABORATIVE WORK IN GROUPS

In my opinion this is the most important learning method of all and I try to use it almost every day. I give them a question or a problem, or a proof relevant to the material currently being studied. First, I give them about 5 minutes to work alone on the problem. Then I break them up into pairs or groups of 3 or 4 to work on the problem together. Since each has already spent time on their own thinking about the problem, this always leads to good interactions in the groups. I move through the class during this group work, observing what is going on and responding to questions. When most groups have answered the question, I reconvene the class as a whole and I ask for a volunteer to present the answer at the board. Then we have a class discussion, hearing answers from various groups. Often I use this discussion as a basis for talking about topics which naturally follow.

In the course evaluations at the end of the semester, 22 out of 23 students liked this method. Although sometimes I let students form groups on their own, I think that it is important for them to be constantly changing groups and working with new people. I do this by having them randomly draw slips of paper with group numbers on it. A quick way to form groups of four in a class of size 28 is to go across the rows counting from one to seven and repeating the count from one to seven four times. Then all the ones form into one group, the twos into another group, etc.

IV.   BOARD PRESENTATIONS

In the previous section I noted that students are given the opportunity to volunteer to make presentations at the board. A problem with this is that some of the same students volunteer all of the time and others never volunteer. To motivate students I give them an automatic full 5 percent toward their grade if they get up and make any presentation. (An example of a positive use of grades as a reward). In appreciation, after each presentation, the class applauds. This gives students a good feeling about going to the board.

Some students want to prepare their presentations before class. To enable this, I let them choose what they'd like to present and I go over it with them ahead of time if they'd like. Even so, there are usually one or two students who will lose 5% of their grade rather than make a board presentation. These are invariably balanced by a number of students who would never make a presentation except under the pressure of it counting toward their grade. Some of these latter students are pleasantly surprised by how well they do and go on to blossom in their self-confidence and readily volunteer to make more presentations.

Helping a student when they get stuck at the board is not so difficult (usually other members of the class are more than ready to help), but dealing with mistakes and conceptual errors is another matter. Ideally other members of the class will pick up the error and correct it. But I've seen serious, blatant errors presented and the entire class sit and say nothing. In this case I usually wait until the presentation is complete before I ask questions to correct the error, because I find that my comments made at the time of the error sometimes fluster the student, and thus prove more of a hindrance than a help. I would appreciate suggestions from others on how to deal with mistakes that students make when presenting at the board. Students have commented that, out of respect, they pay careful attention when a fellow student is making a presentation at the board whereas they feel free to tune-out when the professor is lecturing.

V.   JOURNALS

Each student is required to keep a journal about the course. I ask them to use the right hand pages only and on these they can write anything at all about the current reading assignment and what they are learning about the subject matter. Many math majors object at first to keeping a journal and want more specific guidelines about what to write in their journals. I give them complete freedom to use it in any way they find helpful. If they insist on guidelines, I tell them to write down the main point from each section of the text and tell why they think it is important. Alternatively, they may summarize the material in their own words. In courses in which I give weekly quizzes instead of collecting homework, they may use their journal to work out practice problems.

The blank left-hand pages are reserved for writing questions they'd like to have answered in class or for stating things in the reading that they don't understand. This is the place also for writing about problems which they have been unable to do. I ask them to write down specifically what they need to know in order to do the problem. Often by the time they do that, they figure out the answer for themselves. The questions they write on the left-hand pages should be raised in class. Sometimes when I read the journals, I see questions which they have neglected to raise in class. Usually I don't have time to write the answers in the journal but, if the same question comes up several times, I make a note of it and discuss it in class.

Many students wonder if they will ever learn how to do problems and proofs. Throughout the course I emphasize that one of the best ways to develop such skills is as follows; each time they come to a proof or a worked out problem when reading their text, they should first try to do it themselves. Their journal is ideal for carrying out such work. (They may be motivated to do this because they realize that this work will be noticed when their journals are read). If they can't even get started on the solution or the proof, they should make a note on the left hand page stating what they think they need to know to get started. Then they should look at the first couple of lines in the text to get a start on their own. They should try to finish the write up on their own and refer again to the text only after making a comment on the left page on why they are stuck. When they have finished their own solution or proof, *then* they should read the text write-up. They should use their journal to comment on any differences between their work and the write up in the text. This is too time consuming a process to do always, especially if a student is just beginning to develop mathematical maturity, but it is a great way to develop that elusive mathematical maturity.

To force the students to keep their journals up to date, it is best to collect them regularly, say once a week. During the weekly 20 minute quiz, I have the journals put on my desk, and I quickly check off in my grade book that each is up to date. Then, as the quizzes are turned in, I randomly give each student someone elses journal to take home and read carefully, and write comments on. At the beginning of the next class they return each others journals and I give them some time for discussion. I just read an article [1] about pairing students off and having them keep just one journal between them which they take turns writing in. That seems like an excellent method.

Usually after each hour exam and at the end of the semester, I collect the journals for assessment. Assigning a grade to a journal has its difficulties. I try to make it clear to them that their grade on the journal is not based on what they write but on whether or not they used their journals effectively to help them learn and get involved in the course, a judgement call on my part. During the semester I also try to write comments in their journals which encourage them and show them that I care about what they are doing and recognize their uniqueness.

In the course evaluations at the end of the semester, 10 students liked keeping a journal, 3 disliked, 3 had no opinion, and 7 made other comments. One semester I had five students who have kept journals for me for two consecutive semesters. Four of these students are enthusiastic about keeping a journal and two of these four were top students both semesters. When I asked them if they keep a journal in any course where it is not required they all said "No"!

VI.   WRITING TO LEARN

It should be pointed out to the students that a good way to learn a concept and to understand it is to write it down in their own words. We can profitably incorporate methods which encourage writing into our teaching. One easy way to do this is to assign a problem or a section to be read before coming to class (or have them read it in class) and then ask them to do a "free write" on what they've read. A "free write" means they are free to write anything they want about what they have read. Follow the "free write" by asking for volunteers to read what they have written. This usually leads to a lively discussion. Of course one of the main purposes of journals is to carry out the clarification process that occurs when we write ideas down.

Another writing exercise which is very helpful in getting students focused at the beginning of class is to ask them to write down what they got out of the previous class and/or out of the reading assignment for the present class. They should also write down what they hope to learn in the present class. As usual, a class discussion can be based on this writing exercise. Also it provides helpful feedback to the teacher to collect and read these writings.

VII.   ESSAY QUESTIONS

Another good way to utilize writing to learn is to give essay questions on exams or as homework assignments. Mathematics majors tend to be uncomfortable with essay questions. In the course evaluations at the end of the semester, 12 students liked essay questions, 8 disliked them, and 3 had no opinion. Several said that essay questions should not be on exams.

I use essay questions because they are thought provoking and because of the benefit that comes from putting ones thoughts into writing. I also am uncomfortable with them because I find it difficult to think up essay questions and because I don't know how to grade them. So far I haven't developed any guidelines for the students as to what is wanted in an answer to an essay question and therefore I lack criteria on which to base a grade. Generally if a student puts in a thoughtful effort and shows some understanding, I give them full credit. Even though students find essay questions difficult, they are beneficial to their grade if I'm teaching the course! Some sample essay questions which I have asked are as follows.

1)   In your view what is the most important theorem in hyperbolic geometry? Why?

2)   Write a letter to any one of Gauss, Janos Bolyai, or Lobachevsky. It can be about anything but must relate in part to the material we've covered in this course.

3)   Do you agree or disagree with Gauss in that he did not publish his development of hyperbolic geometry? Present arguments to support the side that you take.

4)   When you come to a proof or a problem when reading your text, why is it always better to try to do it yourself before reading what is written in the book?

5)   Summarize what you learned in the last chapter.

6)   Describe your idea of (A) a mentor, (B) a coach, (C) a teacher, (D) a facilitator. Tell what you like and dislike about each. Which would you prefer to learn with and why?

VIII. GROUP PROJECTS

The importance of projects is widely recognized. In this course I gave two projects and had the students work in pairs. On the first project the students chose their partners by lot, in the second I let them choose their own partners. Although students prefer to choose their own partners, I think that having a partner chosen by lot for one project was good for them and I would do it again.

Following my belief that it is best to give students time to work on something alone before getting together in groups, the first step was to have each person write-up the complete project on their own. Next the partners exchanged their write-ups for peer grading. The graded papers would be returned to the originators. After a few days for assimilation of the grading, the two partners would get together for the first time and work together as a team, clearing up each others questions and trying to arrive at mutual understanding. Then they would sit down individually and write up their final version. Both the original version as peer graded by the partner and the final version were turned in for me to grade. In making up my grade I took particular note of how well they graded their partner's first draft.

In order to foster cooperation between the two partners, on the first project I gave each student the average of their grade and their partner's grade. Some good students had their grades lowered by using this method and they objected strongly. On the second project, I gave each student their own grade independent of their partner's grade.

IX. SELF-EVALUATIONS AND TEACHER EVALUATIONS

In the last week or two of the semester each student turns in a written evaluation of their performance and accomplishment in the course. This self-evaluation counts as 5% of their grade. I make it clear to them, if they make a thoughtful self-evaluation, they will receive their full 5% credit regardless of what grade they think they should get in the course.

I have found reading these self-evaluations to be very interesting, informative, enjoyable and rewarding. A surprising number of students said that they worked harder in this course than any other. Do you suppose that is true or were they trying to impress me? In this regard I was surprised by the number of students who thought they should receive a high grade because of the time and effort they put into the course regardless of their comprehension and understanding and progress.

My own evaluation counts toward another 5% of their grade and is like a class participation and performance grade. It is based on their involvement, initiative, cooperation, attitude, contributions, and the responsibility they took for their own learning. I write this evaluation at the end of their self-evaluation paper before I return that to them. If they have done a good job I use this as an opportunity to express my appreciation of the work they have done. I try to pick out some explicit thing to comment on.

X.   GRADING

Generally I base grades on exams, homework, projects, journals, board presentations, self-evaluations, and teacher-evaluations. In the past I have tried to have class discussions on how much weight should be put on each category. Often it has been difficult to come to a consensus and many students prefer for me to set the weights. Lately what I've been doing is setting certain weights and giving them a range of choices for the others. By a specified date the students must turn in a written statement of the percentages they choose for each category. For example I may give them the following.

5%	Self-evaluation
5%	Teacher evaluation
5% to 10%	Board presentations
5% to 20%	Journal
10% to 30%	Projects
5% to 20%	Homework
30% to 45%	Hour exams
15% to 25%	Final exam

This gives the students some leeway in the way they want to be graded. For example, a student who didn't want to do homework and wanted to be graded mainly on exams could take the minimum of 5% on the homework, and the maximum of 45% on the hour exams and 20% on the final. Other studenta who felt they didn't test well could take the minimum percentages on the exams and take 20% on the homework and 10% on the journal, say. Within bounds, this gives flexibility in how each student is graded.

XI.   STUDENT INPUT

Early on and about every 4 weeks throughout the semester it is important to take time to get some feedback as to how the students think things are going and how they can be improved. I usually ask them to turn in written answers to specific questions such as:

(A)   Should everyone be required to make a presentation at the board? Why or why not?

(B)   Is writing journals worthwhile? Explain.

(C)   Suggest a topic for a group project. Make believe you are the professor and write up an assignment sheet to be passed out describing that project.

(D)   Make at least one suggestion for improving how this course is run.

I read these papers carefully before the next class, tabulate a summary of the responses, and then at the next class, after passing back their papers, I begin by presenting my summary of their responses. We then have a discussion, usually quite lively, and together we decide what changes, if any, we will make in the way we are running the course. The class may be divided on some issue and they may urge a vote. I try to continue the discussion until a consensus is reached, rather than take a vote. If there is quite a bit of pressure to vote on a matter, I lead them to a debate on the question of whether it is better to make decisions by consensus or by voting. This in a math class!

XII.   SUMMARY AND OBSERVATIONS

It is probably best to start slowly by trying one new method at a time, but let the students know at the start that you are going to try some new methods. Begin using one right away and work on that until you are satisfied and then branch out to try another new thing. With each class, at the start of the semester, it is important to spend time getting to know your students individually and to break them into changing groups so that they get to know one another and so that good dynamics develop. Regular feedback methods should be used to keep the class running well.

It takes extra work, dedication and flexibility to bring about change and incorporate new methods. Many students resist at first but in the end they usually come around. When you look back at the end of the semester you will be happy about some of the things you did and unhappy about others. But it is certainly exciting, challenging, and rewarding to become closely involved in a cooperative learning endeavor with your students so that you get to know them and care about them at a deep personal level. Every class meeting is an adventure and you are not sure how it will come out.

Finally I want to acknowledge that many of the ideas and methods I have discussed in this paper I learned at workshops and conferences conducted by the Institute for Writing and Thinking at Bard College. Other meetings and conferences as well as books and papers have also been helpful. Writing Across the Curriculum Workshops at the University of New Hampshire, our bi-weekly seminar in mathematics education, recent meetings of the AMS and MAA and, of course, this particular Conference on History, Geometry and Pedagogy held in honor of Howard Eves have all influenced my efforts to improve my effectiveness as a teacher.

REFERENCE

[1]   Caprio, Mark, "Exchange Journals," *Cooperative Learning*, Vol. 11 No. 4 (July 1991), 30-31.

Calculus in Context: Integrating Calculus Concepts and Finite Mathematics in a First College Course

Stephen J. Turner
Babson College, Babson Park
Wellesley, MA 02157

The introductory calculus course taught at Babson has evolved into a combination of finite mathematics and optimization techniques. The motivational element in topic coverage is provided by illustrations of applications in business. This applications orientation is due to two primary factors. First, almost all of our faculty is involved in consulting and each is encouraged to bring the knowledge so gained into the classroom. Second, one of the reasons students come to Babson is to enjoy our applications orientation. Babson students demand to see business applications in their course content. Our faculty are glad to oblige. One of the popular methodologies is to begin each topic with an applied problem designed to motivate interest in following the theory. The theory is then used to solve the problem and topic coverage expands from this foundation.

COURSE OBJECTIVES

In terms of general developmental course objectives we seek to do the following.

1. **Motivate** by stressing business applications, forcing in-class participation and using examples of an inter-disciplinary character.

We find that learning must be active (versus passive) in order to motivate today's students. If they cannot actively participate they become indifferent resulting in poor class attendance, marginal performance on homework, and other unacceptable behavior. One of our more effective tools used to discourage such unacceptable behavior is the seating plan. Seating plans allow us to initiate student participation fairly and comprehensively early in the course. Students are called by name and asked to contribute to the discussion. Homework problems form the foundation for this participation, but occasional queries dealing with lecture topics are also utilized. Interdisciplinary problems from economics and the functional areas of accounting, finance, marketing, and management always get the attention of our students since these areas are part of our required curriculum.

2. **Inculcate good study habits** by encouraging consistency of effort, regular attendance, good organization of study materials, and the building of a team concept.

We give a quiz every week to encourage a consistent effort at understanding the material throughout the course. Most quizzes are given on Friday. Students are given guidelines for organization of their class notes and homework notebooks. We also coach our students on how to study for exams, giving encouragement and fostering the "class as a team" concept.

THEMES

We utilize the theme concept of course development. Course themes seem to work well when students are only required to take a single course in a given area. When topics are centered on a theme, there is a constant "revisiting" of topics throughout the course. Through such multiple visitations, the themes are conveyed and our educational objectives are realized. We use the following themes.

1.   Linearity and Optimality

The linear theme is accomplished via the topics of linear total cost equations, linear total revenue equations, linear break-even analysis, linear programming, and estimation of exponential function parameters using the log transformation. The optimality theme is accomplished via the topics of linear break-even analysis, linear programming, non-linear programming (classical calculus), and non-linear break-even analysis (response functions used in marketing analyses and integral calculus as applied to income and expense rates).

2.   Marginal Analysis

A key element in our concept of applied calculus is the interpretation of the derivative as an instantaneous marginal cost or revenue (demand or supply in an economic context). Therefore, the "finite" interpretation of marginal cost and revenue is introduced early in the course. Piecewise-linear total-cost functions provide an excellent tool for this introduction since the marginal cost changes at each intersection point of the component lines. Reenforcement is provided in linear programming through sensitivity analysis. One simply notes that the marginal cost or profit per unit produced remains constant within certain ranges which leave the variables in the solution unchanged. (However, the values of the variables in the solution may change.) The marginal concept is revisited a final time during our coverage of the derivative.

3.   Computer Assisted Solution Techniques

We place heavy reliance on computer assisted solution techniques. Computer algebra system software is utilized in course sections populated by students with weak algebra skills. Linear programming software is used for solving linear programming problems. Spreadsheets and system programs are used for solving financial problems. Graphics software is used throughout the course. We consider these software packages and system programs to be assets in the learning process. Students find them appealing and, therefore, they generate interest in the subject in general. Their use allows us to concentrate on concepts, problem formulation, interpretation of results, and the importance of quantitative approaches in assisting the management decision process.

SEQUENCE OF TOPICS

Our topic sequence is as follows.

1. Linear total cost and total revenue functions.
   a. Economic interpretation of slope (variable cost) and intercept (fixed cost).
   b. Piecewise linear total cost functions.
   c. Quantity discount model (intercept has no economic interpretation).
   d. Marginal analysis.

2. Break-even analysis.
   a. Production (how many to make and sell; how many to sell to cover costs).
   b. Accounting (how many to order and sell; how many to sell to cover size of order).
   c. Marketing (how to offer equivalent options to consumers).

3. Linear programming.
   a. Financial portfolio composition, aggregate planning in manufacturing.
   b. Linear programming software package.
   c. Sensitivity analysis and the marginal concept.

4. Exponential growth.
   a. AIDS (number of cases, medical care costs).
   b. Forecasting via the natural log transformation (this coverage is a bridge between our calculus and statistics course as the concept and role of linear regression is introduced).

5. Financial modeling.
   a. Compound interest (interest rates – nominal, per conversion period, and yield).
   b. Future value, present value, term, and discount rate.
   c. Ordinary annuities (mortgages, sinking funds, and municipal bonds).
   d. Cash flow analysis (effect of delayed payments, perpetual annuities, maintenance annuities).
   e. Computer assisted techniques (spreadsheets and other system programs utilized).

6. Applications involving finding maxima and minima of nonlinear curves. (Bridge is to discuss the problem of handling nonlinear effects in the context of optimization problems covered earlier in linear programming – that is, nonlinear objective functions or nonlinear constraints wherein the optimal solution is still on the edge of the feasible region but we know not where.)
   a. The concept of a limit (emphasis on "L'Hospital-type" limit problems which force students to think in terms of the true concept of a limit).
   b. Revisit marginal concept to develop Newton difference quotient methodology.
   c. Inventory and production model problems (minimum cost of ordering and holding inventory; optimal production run size).
   d. Least-squares curve fitting.

7.   Applications involving calculation of areas under curves.
      a.   Profit as a function of income and expense rates.
      b.   Demand and supply curves and calculation of producer and consumer surplus.
      c.   Numerical integration (another bridge topic since the standard normal probability
              distribution is used as one of the examples).

Throughout the course our generic methodology is to introduce each topic using an illustration of a simple but realistic "pseudo" application. The purpose is to motivate interest in the general topic without intimidating the students with details. This is followed by coverage of the mechanics of the method in a "concept" model-building context which relies on simple but comprehensive cases. (For example, in linear programming we cover two variable problems with emphasis on determination of objective function and constraints, the notion of a convex set of feasible solutions, iso-profit lines, and multiple optima.) A key element in our methodology is an emphasis on interpretation of results (both oral and written communication skills). The idea is to accomplish an effective transfer of information to a naive third party (a manager, for instance, who has no knowledge of the mathematical methodology or terminology employed in obtaining the problem solution). This emphasis also allows us to underline the fact that computer output must be interpreted and communicated to be of value. The following example is an illustration of an applied problem which connects simple break-even analysis, two-dimensional break-even analysis, linear programming, and interpretation of computer output.

EXAMPLE

In 1990, the Milford, New Hampshire Earth Day Committee ordered some three-color T-shirts to sell as a method of defraying Earth Day celebration costs. The T-shirt cost was $25 per screen (one screen per color) and $6.25 per T-shirt ordered. The Committee's initial order was for five and one-half dozen.

1.   The selling price to the general public was $10 per T-shirt. How many of the 66 T-shirts ordered did they need to sell in order to break-even?
      Solution:
              Let p denote the number of the 66 T-shirts sold to the general public. The revenue and cost equations are, respectively, $R(p) = 10p$ and $C(p) = 487.5$, so that $p = 48.75$ when $R(p) = C(p)$.

2.   The selling price to Committee members was $8 per T-shirt. Fourteen T-shirts were sold to Committee members prior to any sales to the general public. How many of the remaining 52 T-shirts did the Committee need to sell to the general public in order to break even?
      Solution:
              Since 14 T-shirts have been sold to Committee members, the revenue and cost equations are, respectively, $R(p) = 10p + 112$ and $C(p) = 487.5$, so that $p = 37.55$ when $R(p) = C(p)$.

3.   When two different selling prices exist, the revenue and cost equations each have two variables, one variable corresponding with each price. In such a case there is a break-even

line rather than a break-even point. Find the break-even line assuming that 66 T-shirts have been ordered and 14 have been sold to Committee members.

Solution:

Let c denote the number of the 66 T-shirts sold to Committee members. The revenue and cost equations are, respectively, $R(c,p) = 8c + 10p$ and $C(c,p) = 487.5$, so that $8c + 10p = 487.5$ when $R(c,p) = C(c,p)$.

4.   Note that not all points on the break-even line are feasible. For instance, c must be at least 14 and $c + p$ cannot exceed 66. These constraints determine a set of feasible combinations of values for c and p. The break-even line partitions this set of feasible solutions into a "loss" region and a "profit" region. In fact, since profit is equal to revenue minus cost, one legitimate question is what combination of c and p maximizes $8c + 10p - 487.5$?

Solution:

Since the profit for each sale to the general public is greater than that for each sale to Committee members, the optimal solution is, obviously, $c = 14$ and $p = 52$. However, it is important to note that the problem can be formulated using a linear programming model as follows: maximize $8c + 10p - 487.5$ subject to the constraints $c + p \le 66$ and $c \ge 14$. Additional information can be obtained by interpreting the LINDO (Linear INteractive and Discrete Optimizer, developed by Linus Schrage at the Graduate School of Business, University of Chicago) software package output. The package will not allow a constant to be used in the expression to be optimized. Thus, the following LINDO formulation includes a dummy variable, D, which accomplishes the desired result.

```
MAX   8 C + 10 P - 487.5 D
SUBJECT TO
      2)    C + P <= 66
      3)    C >= 14
      4)    D = 1
END
```

The software gives the solution as follows.

```
LP OPTIMUM FOUND AT STEP 4
        OBJECTIVE FUNCTION VALUE
  1)    144.500000
```

VARIABLE	VALUE	REDUCED COST
C	14.000000	.000000
P	52.000000	.000000
D	1.000000	.000000

ROW	SLACK OR SURPLUS	DUAL PRICES
2)	.000000	10.000000
3)	.000000	-2.000000
4)	.000000	-487.000000

Here is an interpretation of this LINDO output for the purpose of informing management. The maximum amount of money that can be added to the celebration coffers is $144.50 unless we place another order above and beyond the initial order of 66 T-shirts. The $144.50 amount will result only if all of the remaining 52 T-shirts are sold to the general public. The marginal increase in the maximum profit for each additional T-shirt ordered and sold is $10.00. The marginal decrease in the maximum profit for each additional T-shirt sold to Committee members is $2.00.

5.   Total costs for the celebration were estimated at $800 (over and above the cost of the T-shirts). Contributions were estimated at $500. Since the solution to the above problem is $c = 14$ and $p = 52$, and the corresponding profit is $144.50, the Committee must order an additional number of T-shirts to cover the $300 gap. How many more T-shirts should be ordered?

Solution:

The Committee has no guarantees on sales and so needs to be as conservative as possible. If the additional number of T-shirts ordered is "small" then they can assume that all T-shirts ordered will be sold and the solution can be found from the following linear programming formulation: minimize $c + p - 66$ subject to the constraints $c + p \geq 66$, $c \geq 14$, and $8c + 10p - (6.25c + 6.25p + 75) \geq 300$. (Note that $c + p - 66$ is the minimum order size required.) The LINDO formulation is as follows.

```
MIN  C + P - 66 D
SUBJECT TO
      2)    C + P >= 66
      3)    C >= 14
      4)    1.75 C + 3.75 P >= 375
      5)    D = 1
END
```

The software gives the solution as follows.

```
LP OPTIMUM FOUND AT STEP 4
        OBJECTIVE FUNCTION VALUE
  1)      41.4666700
```

VARIABLE	VALUE	REDUCED COST
C	14.000000	.000000
P	93.466670	.000000
D	1.000000	.000000

ROW	SLACK OR SURPLUS	DUAL PRICES
2)	41.466670	.000000
3)	.000000	-.533333
4)	.000000	-.266667
5)	.000000	66.000000

Here is an interpretation of this LINDO output for the purpose of informing management. The Committee must order at least 42 additional T-shirts in order to cover celebration costs. Each of the remaining 52 T-shirts in the current order and all of the additional 42 T-shirts must be sold to the general public. The marginal increase in the minimum additional order size is .5333 T-shirts for each additional sale to a Committee member. Thus, if 15 additional T-shirts were sold to Committee members (bringing total sales to Committee members to 29) then the minimum additional order size would increase to about 49. The marginal increase in the minimum additional order size is .2667 T-shirts for each dollar added to celebration costs. Thus, if a story-teller is hired at a cost of $50.00, then the minimum additional order size would increase to about 55. If the supplier restricts order sizes to be in half-dozens, then the research team will need to reassess the situation and return with another report.

Howard Eves – Some Recollections

I want to share some personal recollections about Howard Eves. I believe such recollections are appropriate since this volume is a celebration of his eightieth birthday and of the contributions he has made to mathematics during his lifetime. Many nice things have been said and stories told about Howard during the conference. My story centers on what I consider to be Howard's most unusual characteristic.

Howard was my teacher for three courses I took at the University of Maine, and I used to visit him in his office on a regular basis. Perhaps you have forgotten how it was to be a student and to visit a professor in his office. I became somewhat of an expert in the area of faculty office layout and design during my years as a student at the University of Maine. Let me demonstrate by presenting a couple of examples of strategies for office layout that I actually encountered.

I call the first of these designs "the squeeze play." The design applies to a rectangular space and the arrangement is as shown below.

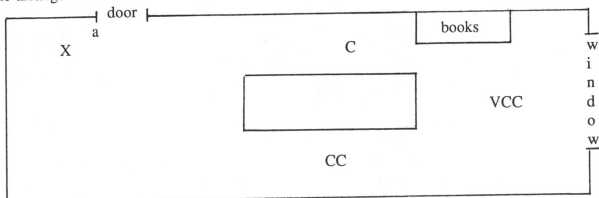

The letters denote the following:

    C,   typical "office" or "visitor" chair (hard and uncomfortable to sit in for extended periods of time);

    CC,   an office desk chair (comfortable with arm rests);

VCC,   a very comfortable, padded, "parlor-type" chair;
  X,   coat rack;
  a,   door hinge.

Note the subtle but effective elements of this design. First, the comfortable chair, CC, is for the professor. The space allotted this chair is large to allow for "leaning back," leg room, etc. Now the room is not very wide. Once the desk is accounted for, this leaves a space just big enough to wedge the visitor's chair between the opposite wall and the desk. A student visitor is forced into the fetal position during a conference. The bookcase blocks access to the very comfortable chair situated in front of the window. Thus, the student must sit uncomfortably while viewing a very comfortable impossibility. This is very effective in discouraging student visits and in quick dispatch of those unfortunates who do arrive. A note of interest to professors who teach in northern climates; the location of the coat rack. Since the door is hinged at "a", an arriving student opens the door and hides the coat rack at the same time. The student never knows the coat rack is available and so sits with coat on and begins to sweat. The student quickly resolves that most of the information to be gained from the conference was not that important after all.

The second of these designs is called "the king's court" and is shown below.

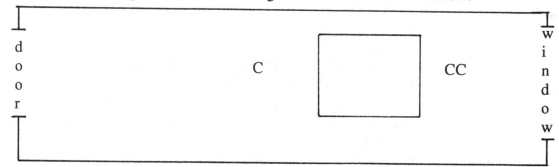

The idea behind this design is to create an impression of a long approach to a throne. If the walls along the long dimension are kept uncluttered, that impression is practically guaranteed. This design is particularly effective on sunny days as the professor appears to have an aura about his or her upper torso. Surely any utterance from the "throne" must be cast in stone. Note that if office hours are timed for the hour of the day when the glare is at its peak, then students would be squinting. No student would put up with squinting for very long.

Other generic strategies include installing floor to ceiling book shelves completely stocked. The impression that you have assimilated all the knowledge therein makes a student humble. It also helps to file papers everywhere, especially on visitor's chairs. Not only will the student have no place to sit, but he or she will also know that you are far too productive and busy to spend a significant amount of time in consultation.

These strategies are presented in stark contrast to the design of Howard's office which is shown below.

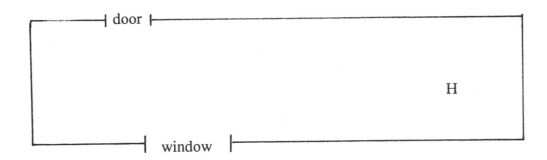

The "H" denotes Howard. The only thing in Howard's office was Howard! No books on the shelves and no papers anywhere. The effect was so startling you almost fell as versus walked into Howard's office. It was clean as a whistle, just as though no one was assigned to it. I know it was a small office as it didn't create an echo.

I am exaggerating. There was a desk with one book on the top, a filing cabinet, and two chairs. The book was the text Howard was teaching from that day. Of course he had his famous briefcase. There was nothing else in the office than these. How remarkable it was to talk mathematics in that office. If you asked a question requiring written elaboration on ANY mathematical topic (topology, abstract algebra, number theory, geometry, history, etc.) the ritual was always the same. A pencil was retrieved from the middle drawer and the drawer was closed; a white unlined piece of paper was selected from the upper right drawer and the drawer was closed. A perfectly executed explanation would follow. Never a mistake, never a hesitation, never a consultation with another faculty member, and, obviously, never a reference to a book or paper. Howard never made a mistake in class either! The only time he used the eraser was during his pre-class board-cleaning ritual and to make room for his next immaculate conception.

Attending Howard's class was like going to the theater. From the start you were captured by the consummate artistry of an effortless exposition punctuated by stories of the great masters. Howard's artistic abilities were second to none. He drew perfect circles free-hand. The figures he put on the blackboard could have served perfectly, without alteration, in any textbook. Any of these abilities alone would have made him an outstanding teacher. Together, they made him unparalleled.

When I go to the great university in the sky, my first stop will be at the registrar's office where I'll run my finger down the course listings until I come to the name "Howard Eves." I know he will be on the list and I can hear his greeting now; "Why,..if it isn't the pastor. Gee,..it's so nice to see you again. I hoped we could get together, you see I've come across the most remarkable result...."